The Essential Guide

Edited by

CHRISTIE WILCOX

BETHANY BROOKSHIRE

JASON G. GOLDMAN

Yale

UNIVERSITY PRESS

New Haven and London

Published with assistance from the foundation established in memory of William McKean Brown.

The authors gratefully acknowledge the assistance of the Alfred P. Sloan Foundation and the National Association of Science Writers. This project was funded in part by a grant from the National Association of Science Writers. Reference to any specific commercial product, process, or service does not necessarily constitute or imply its endorsement of or recommendation by the National Association of Science Writers, and any views and opinions expressed herein do not necessarily reflect those of the National Association of Science Writers.

Yale University Press books may be purchased in quantity for educational, business, or promotional use. For information, please e-mail sales.press@yale.edu (U.S. office) or sales@yaleup.co.uk (U.K. office).

Designed by Sonia L. Shannon
Set in Scala type by Integrated Publishing Solutions
Printed in the United States of America.

Library of Congress Control Number: 2015946024
ISBN 978-0-300-19755-6 (paperback)

A catalogue record for this book is available from the British Library.

This paper meets the requirements of ANSI/NISO Z39.48-1992 (Permanence of Paper).

10 9 8 7 6 5 4 3 2 1

Science Blogging

Contents

Preface

Why are you here?

We don't mean that in the existential sense. But what are you doing, right now, with this book in your hands? (Or more likely with this book displayed on some electronic device?) What is it that you want to know about science blogging?

Online science communication has come a long way from the early days of LiveJournal and Usenet. Bloggers are no longer sending messages in bottles with only blogrolls and hyperlinks to keep us connected, each of us in our own little far-flung corner of the Internet. Platforms such as Twitter and Facebook may not feel all that new, but they're revolutionary tools that have allowed us not just to interact with each other but also to reach wider and more diverse audiences. Many of us are now clustered together in official networks, under the umbrella of large, mainstream magazines or newspapers. Some independent blogs have grown into high-traffic sites, destinations unto themselves. Science bloggers are no longer limited to blog posts but are also writing books, recording podcasts, and uploading videos to YouTube. For many, science communication is a viable career.

When the three of us started blogging, the crowd was smaller. There were fewer science blogs, which meant it was easier to distinguish ourselves from other bloggers. It was easy to find our voices and to make ourselves heard. It was a time when someone sufficiently motivated could read just about every new blog post written about science each day.

That is no longer the reality. Today breaking in to online science communication can seem almost impossible. It can seem like there are dozens of new science blogs—not to mention YouTube channels and podcasts—popping up each day. But the Internet is still very much a frontier for science communicators. It's the Wild West. Each time the scene threatens to become too settled, someone or something new arrives, keeping us all on our toes. The struggle was once to become heard at all; now the struggle is to remain relevant.

Maybe you're reading this book because you wish to be that someone new. To disrupt the status quo. Or perhaps you're here to get new ideas. Maybe you're here to get your blog to the next level, to transition from blogging as a hobby to blogging as a career. Or maybe you're here to figure out just where your voice fits in the online world.

No matter your goals, this book is here to help. We've brought twenty-seven of the most successful, insightful online science communicators together to share with you what their years of experience have taught them. All of their expertise is as current as we could make it; as of this writing all of the online references you'll see are up-to-date and available.

But you don't need to read this book cover to cover to learn what you need to know. Depending on your goals, there are different chapters, different paths through this book that will help you on your way. Here are just a few ideas.

Science Blogging 101

Maybe you are getting started as a blogger and need ideas about how to proceed. In that case, you might start with Chapter 3, to learn the basics of setting up a science blog from Khalil A. Cassimally. Make sure you use and display images and artwork responsibly by checking out Chapter 4 by science artist Glendon Mellow. Then you might check out Chapters 7 and 8 by Danielle N. Lee and Zen Faulkes. They present two different views on science blogging, one from within an official blog network, and one at a personal, independent site.

As you get in gear, check out Chapter 1 by Christie Wilcox to remain mindful of why you're blogging and what you want to achieve. And don't forget that, though it is still young, science blogging has a rich history. See Carl Zimmer's history of our online community in Chapter 2.

Once you have your blog up and running, it's time to think about how you might best communicate your science to the wider world. It doesn't all have to be long essays. Joe Hanson discusses how to be effective by being brief, in Chapter 10. And in Chapter 22, Rhett Allain covers some of the specific challenges that might arise when tackling the hard sciences.

Telling Your Own Story

No one starting as a science communicator wants to get lost in the crowd. We all have different stories to tell, different angles we can use to communicate. But sometimes it can be difficult to find your own unique voice.

Your approach to science blogging might be influenced by your life experiences. In Chapter 11, Ben Lillie talks about using your personal experience to drive your works. Many bloggers are also heavily influenced by their identity. Alberto Roca offers thoughts on blogging as a minority-group member in Chapter 12, while Kate Clancy offers advice to other female bloggers in Chapter 13.

Many people come to science blogging through their careers.

Colin Schultz writes about the benefits of blogging as an early career journalist in Chapter 14, while Marie-Claire Shanahan covers blogging as an educator in Chapter 16. Karl M. Bates considers blogging from inside the ivory tower as a public information officer in Chapter 15. For researchers who are blogging as a form of scientific outreach, Jason G. Goldman covers blogging as a graduate student in Chapter 17, while Greg Gbur takes it to the tenure track in Chapter 18.

While many fine blogs exist to explain concepts and cover research papers, not all of them need to conform to this standard. Scientist Travis Saunders and science writer Peter Janiszewski describe how they use their blog to cover scientific conferences in Chapter 21.

I Have a Blog. Now What?

If you're already an experienced denizen of the online world, this book still has plenty to offer. Maybe you've been a small fish in the big science pond too long, unsure about how to find readers. Ed Yong offers insights on how to find your audience in Chapter 5. Many science bloggers are held back by their discomfort with self-promotion, but Liz Neeley will show you, in Chapter 20, that you have nothing to fear. Some bloggers may want to try moving beyond using words alone. To use interactive tools so that your readers will become participants rather than passive viewers, check out what Rose Eveleth suggests in Chapter 9.

Many science bloggers and communicators may think they are reaching the audience they want, but aren't really sure how to look at their metrics. How do they know if anyone is really listening? Matt Shipman offers important insights on metrics in Chapter 19.

Writing on the Internet can be daunting. In Chapter 23, Emily Willingham covers how, and why, we might choose to write about controversial topics, and Melanie Tannenbaum gives some tips for countering trolls in Chapter 24. In Chapter 6, Janet D. Stemwedel shows how carefully we must consider ethical issues as we share information on the Internet.

Many blogs are written simply for the love of science and science communication. But career science communicators also need to pay the bills. Bethany Brookshire discusses getting paid for blogging in Chapter 25. And in the final chapter, Chapter 26, Brian Switek explains how your blog can serve as a sort of "writing laboratory" to help you develop ideas for the ultimate in long-form writing, a book.

By bringing together some of the most experienced voices from around the science blogosphere, we hope this book will have something to teach everyone. Whether you're just getting started, have some blog posts under your belt, or are looking for fresh inspiration, you are not alone. The science communication community may seem overwhelming, but it's friendly. Dive in and show us what you can do. Seriously. Tweet us and show us your stuff. And use our hashtag, #SciBlogGuide, and find us online at http://www.theopennotebook.com/science-blogging-guide.

Bethany Brookshire (@scicurious), Washington, D.C.
Jason G. Goldman (@jgold85), Los Angeles
Christie Wilcox (@nerdychristie), Honolulu

Science

Blogging

1

To Blog or Not to Blog

CHRISTIE WILCOX

What does a budding scientist, science communicator, or science writer have to gain from writing about science on the Internet? What are the benefits of getting started in social media? Popular blogger Christie Wilcox of Discover Magazine Blogs *takes you through the many reasons why you, yes you, should start communicating about science on the Internet.*

You picked up this book, so you must be at least a little curious about starting a science blog. Or maybe you already have one but could use a little validation. You want to know *why* you should write a science blog. After all, aren't there a million blogs out there?[1] Isn't the Internet bursting at the seams already? Why should you, a busy academic/scientist/journalist/writer/public information officer/insert-your-title-here take the time to write online (especially if it's unpaid at first)? Why should you bother with this often-maligned medium, when there are journal articles or features to be written? Really—why should you, of all people in the world, be blogging?

While I can share my personal experiences and give you a hundred reasons to blog, ultimately they boil down to two philosophical principles: altruism and narcissism. If you ultimately decide to blog, it will be either for yourself, for the good of others, or a bit of both.

Make the World Better for Science: The Altruistic View

If you like to think of yourself as a giving person, then blogging is definitely right up your alley. What better way to share your passion and love of science with the rest of the world? And the truth is, the rest of the world needs it.

Now, more than ever, science is fundamentally intertwined with national and international political issues. Our climate is changing. Animals and plants are going extinct at an alarming rate. Life-saving technologies like vaccines are denigrated and misrepresented. Every day technologies advance in ways that are rarely explained well to the rest of society. To make informed decisions on a wide range of political issues, the people of the world need to understand the science behind the most hotly debated topics. But to do that, they need interpreters who speak the lingo, who can take jargon-filled research and put it into terms that anyone can understand.

Nowhere is this more true than in the United States, where former Senate majority leader Trent Lott can call his four years of science and math in high school "a waste of my time and a waste of my teachers' time" and receive roaring applause.[2] The only way to change the negative attitude toward science is to show people why they should care. To do that, we have to show people how incredible, important, and intriguing science really is.

But you already know that. You've already had moments where your passion just bubbled out of you uncontrollably, and you saw the spark in someone's eyes when you told them something really, *really* cool. Maybe you explained how the Higgs Boson particle works or got into a conversation about the ballistic penises of male ducks. Somehow you found yourself in that place of authority where you were sharing with others something new and fascinating, and you changed how they think about the world—just for a moment, or perhaps forever. You inspired them. You want to do it again. And you can use a science blog to do it.

There's a reason that major grant agencies like the National Science Foundation require outreach and communication from the scientists they fund. Reaching out and sharing science is a moral responsibility for those who "get it." As the American Psychological Association's David Ballard says, "We have an obligation to be out there in public because there is nobody better informed or more expert."[3]

So why not attend more conferences or go talk to students in their classrooms instead? Because we live in a digital age where ten-year-olds carry smart phones and information is never more than a Google search away. More than half of Americans say they "talk" to people online more than they do in real life.[4] As social media platforms continue to grow exponentially, people are turning more and more to online avenues for connection and communication. If we want to be involved in the conversations about science, we have to be online as well. We need to be found in search results, and get real, accurate science into online conversations.

Most importantly, online avenues target the everyday adult. While we can improve education in public school and try to fight the battle in the next generation, we have to go beyond to really shift our culture. Anyone born before 1980 (and some born after) didn't learn about stem cells in high school. They aren't going to be taking a traditional class to better understand climate change or the causes of autism. They will learn about and understand these issues better only if they have access to content that explains them clearly.

Blogs reach out far beyond even the most gregarious person. I've had blog posts translated into Chinese, Romanian, and French. Commenters come from around the world to weigh in on the science I discuss on my blog. And because it's the Internet, what is written on a blog doesn't just stay on that blog; a wide variety of media outlets and other major traffic sources link to it as well. "Just simply by having a blog," says Travis Saunders of *Obesity Panacea* (http://blogs.plos.org/obesitypanacea), "we've been able then to go and get our message out to literally hundreds of thousands or even millions of people through these other much, much larger engines."[5]

I've seen firsthand the immense reach of blogs. When I wrote about DNA fingerprinting to explain how Osama bin Laden's body was identified, more than eighty thousand people read the post on my blog over the next couple days, and it was linked by PBS *NOVA,* NPR, *Nature, Discovery,* and a suite of mainstream news organizations, not to mention other blogs. It took only thirty minutes for me to explain the science behind something I do every day, yet millions of people learned about genetic fingerprinting and were able to explain to their networks how we knew bin Laden was dead.

That's the point, isn't it? To get people talking about science, thinking about science, *caring* about science. To help people find science in the everyday.

Science blogging is truly a noble pursuit because it seeks to inform and excite others. It's all about injecting your personality, your passions, and your reasons for loving science into online content that educates and inspires. The ultimate aim is to change the world—a lofty goal with all the feel-good, heartwarming hope you could ever want in an activity.

It Really Is All about Me: The Narcissistic View

The simple truth is that no matter how much good we want to do for the world, we are all limited. We are, as they say, only human. We have jobs that need to be done, money that needs to be made, and personal lives to attend to. So why should you make time in your busy schedule for blogging? Because ultimately, you're the one who reaps the most rewards from it.

Let's start with the most immediate benefit: exposure, or as marketing professionals refer to it, "personal branding." A long time ago, when you applied for a job or met a new person, they only had one thing to judge you by: what you told them. Now, in less than a minute, a potential employer or colleague can learn a lot about you. If you had Googled my name before 2006, for example, the top result would have been a quote my eighth-grade self gave to my middle school newsletter.[6] The Internet never forgets, and you can either lament that fact or do something about it. Blogging is content over which you have 100 percent control. That means when someone searches your name and finds your blog, they are seeing what you want them to see—your words, your thoughts, evidence of your skills and expertise.

Nowadays, it's more likely that a lack of web presence will damage you as you apply to new jobs. Just ask danah boyd, an assistant professor in media, culture, and communications at New York University and a visiting researcher at Harvard Law School. "There is no doubt that all faculty searches include a Google search," writes boyd.[7] "One of the things I hear most frequently about a new hire is how disturbing it is that he doesn't have a web presence. Something must be wrong, right?"

The best part of having your own blog is that these potential employers, colleagues, or whomever will get to see the *best* you. Instead of being a name and a résumé, you'll be a person—and you've already begun charming them, even if they haven't met you face to face. In that way, blogging provides another benefit: it's like regular networking, but without the pesky limitations of location and timing. Blogs are inherently interactive platforms. With comment threads and the ability to link around the world, they're all about conversations. Instead of rubbing elbows with a handful of people at a small, in-person function, you're chatting with thousands of people from all walks of life, any of whom might become an important contact later on. I know firsthand that this can occur: I first met one of my collaborators on my dissertation thanks to blogging.

Others have similar stories. Bertalan Mesko of scienceroll.com feels that "blogging and Twitter don't just help me in my research but totally changed the way I interact with other researchers and collaborators." Similarly, John Fossella (who blogs at genes2brains2mind2me.com) has found that blogging has expanded his scientific network. "Instead of getting feedback from the same handful of folks I regularly see in the lab, I'm getting comments and new ideas from folks who I used to work with 5, 10 and even 20 years ago, not to mention new folks who I've struck up online interactions with."[8]

"Science blogging literally changed my life," explains Australian science writer Bec Crew, who didn't know how to get started when she graduated with degrees in arts and media. Initially, she started blogging to satisfy her need to write while working an office job to cover the bills. As she gained attention for her posts, opportunities opened up, and Crew credits blogging with launching her career. "I was completely unqualified for the position I applied for at one of Australia's few science magazines, *COSMOS*," she explains, "but there was no questioning my enthusiasm for science communication, which helped me get the job." She was even approached to write her first book, *Zombie Tits, Astronaut Fish, and Other Weird Animals* (which came out in October 2012), through her blog. When it comes to blogging, Crew says the time put in is 100 percent worth it. "It's proof of your commitment to the industry, which is especially handy if you haven't had the opportunity to work in it professionally yet." More importantly, while it's easy to say you're a good writer, hard-working, or committed, showing it is much harder to do—and so means a great deal more. Blog posts can serve as writing samples to show editors, and because they're online, an interested editor will have an immediate, easy way to contact you. As Crew writes, "What will set you apart is the fact that you've been writing about science in your own time, and training yourself to be better at it; because you love it and you think it's important."[9]

This is especially true for the scientist blogger. Science is a labor of love. You do what you do because you think it matters, and you publish your research because you think it's worth talking about. What better way to make sure your research is talked about than to start the conversation yourself? Multiple studies have shown that media attention can positively influence paper citations.[10] This is especially true because, as U.K.-based geneticist Daniel MacArthur has noted, "a fairly hefty proportion of the readership of most science blogs consists of other scientists, so having your work disseminated in these forums . . . increases your profile within the scientific community, promotes thoughtful discussion of your work and can lead to opportunities for collaboration."[11] And if your research is already being talked about widely, you *definitely* want to be blogging. As GrrlScientist explains in a post about scientists blogging, "A blog can be used to rapidly correct errors in mainstream media reporting, and to highlight the value of one's findings while doing so. But perhaps most important, a blog provides scientists with a public platform where they can defend their research from misuse or misrepresentation by politicians and corporations that seek to abuse scientific data to bolster their agendas."[12] As the #arseniclife scandal made blatantly clear, your research is fair game for other science bloggers. When NASA-funded scientists published the shocking finding that some bacteria can replace phosphorus with arsenic, they found out the hard way that in this Internet age, scientists will not just challenge your results academically, they'll also do it online in full view of the public.[13] "Savvy scientists must increasingly engage with blogs and social media," explains Paul Knoepfler, professor of cell biology at the University of California, Davis School of Medicine, in a comment for *Nature.* "Even if you choose not to blog, you can certainly expect your papers and ideas will increasingly be blogged about. So there it is—blog or be blogged."[14]

For all writers of all kinds, from journalists to novelists, there is no better way to get yourself and your work out there than to write more. A blog is a writing laboratory where you can experiment with types of content and see what works best and what doesn't. You can play with images, videos, and all sorts of multimedia. It requires commitment, which means you're putting words to the page, showing potential employers that you have the dedication and ability to produce content. Blogging also keeps you keyed in to the most recent and relevant scientific discourse, and allows you to interact with other writers and the scientists whose work you write about. You'll sit at the same table with some of the most well-respected science communicators out there and gain insight into what they do and how they do it.

Science journalist Carl Zimmer has found that blogging allows him to expand his topic range and elaborate on new ideas. "I blog about things that I find very cool but that I won't be able to turn into an article someone will pay me to write," says Zimmer. "Very often, I will mine these posts for my books, and I sometimes even manage to produce articles on topics I first visited on my blog."[15]

If nothing else, blogging helps develop essential skills. "A wonderful side effect," says Ph.D. student Drew Conway, "is that the overall quality of your work will also increase, as you become a better writer, researcher and conveyer of complex ideas."[16]

National Geographic blogger Ed Yong reminds us that for a journalist, blogging is a great form of practice. "When I write for my blog, I do so in exactly the same way as I would for a mainstream organization. I ask whether stories are worth telling. I interview and quote people. I write in plain English. I provide context. I fact-check . . . a lot. I do not use press releases, much less copy them."[17]

By blogging you practice writing, cohesive thinking, effective communication, and web skills like HTML programming, skills you will use no matter what future career you find. "It's really an opportunity to work on your writing and presenting skills," says Saunders. "I found that having an excuse to write every day, trying to distill research down into lay terms . . . gave me a lot more confidence in my writing ability and also my confidence in presenting."[18] As every writer knows, the best way to improve as a writer is to write—and blogging not only nudges writers to write more regularly; it also provides wiggle room to explore different narrative structures and writing styles.

While blogging might seem a selfless act at first, it opens the door to real career-enhancing opportunities, whether that means broadening your professional network, increasing your exposure, or simply making you more marketable through new and enhanced skills. You can even end up making money off of it, though I wouldn't recommend getting into science blogging for the cash. Blog because you like to communicate, and because you have a passion for scientific topics that need someone like you to convey them. Blog to gain exposure and network, and to expand your career.

The Win-Win of Science Blogging

Scientist bloggers can gain a wider audience for their research, network with other scientists they might never have otherwise met, and establish their name as experts in their fields. Bloggers who focus on science, whether they're scientists, journalists, writers, or simply enthusiasts, can use their blog for self-promotion, draw in larger audiences, practice important skills, and try out new ideas and media types. Meanwhile all science bloggers benefit from being involved in a conversation and receiving feedback and ideas from a much bigger audience than they would reach with traditional outlets. And they get to do all of this while doing the important work of sharing what they love with the world, shifting negative cultural attitudes toward science, and combating pseudoscience and misinformation.

So you, enthusiastic science-y person, why aren't you blogging already? If you're feeling inspired but still unsure of the next best steps, the rest of this book will help you start with advice and how-tos from the best in the business, so you can begin to reap the many rewards. And if you are already blogging, this book has information for you as well: their insider know-how will help you take your blog to the next level, so you can reach whatever goals you have set for your little corner of the Internet.

CHRISTIE WILCOX is a science writer and social media specialist. She is a postdoctoral researcher at the University of Hawaii, Manoa, where she uses molecular biology techniques to study venoms. She is also a writer at *Discover Magazine Blogs,* where she authors the blog *Science Sushi.*

Christie Wilcox is based in Honolulu. Find her on her website at christiewilcox.com or follow her on Twitter, @nerdychristie.

Notes

1. Actually, there are hundreds of millions of blogs—over 70 million on WordPress alone.

2. Jeffery H. Toney, "Physics: 'A Waste of Time?,'" *Huffington Post,* July 6, 2011,

http://www.huffingtonpost.com/dr-jeffrey-h-toney/physics-a-waste-of-time_b_845184.html.

3. Anna Miller, "You: The Brand," gradPSYCH, November 2012, http://www.apa.org/gradpsych/2012/11/you.aspx.

4. Alex Trimpe, "Google Think Insights," *Think Quarterly* (April 2011).

5. From a video by Saunders posted in Peter Janiszewski, "Social Media for Scientists: A Lecture," *Science of Blogging,* October 12, 2002, http://scienceofblogging.com/social-media-for-scientists-a-lecture.

6. Seriously. See my quotation at http://bit.ly/SsbLuE.

7. danah boyd, "Bloggers Need Not Apply: Maintaining the Status Quo in Academia," *Apophenia,* July 11, 2005, http://www.zephoria.org/thoughts/archives/2005/07/11/bloggers_need_not_apply_maintaining_status_quo_in_academia.html.

8. Quoted in Hsien-Hsien Lei, "Scientists and Social Networking: A Primer (Part 2)," *HUGO Matters,* February 17, 2010, http://www.hugo-international.org/blog/?p=145.

9. Bec Crew, "How Science Blogging Can Lead to a Science Writing Career," *Scitable* (blog), *Nature,* September 11, 2012, http://www.nature.com/scitable/blog/scholarcast/how_science_blogging_can_lead.

10. David P. Phillips et al., "Importance of the Lay Press in the Transmission of Medical Knowledge to the Scientific Community," *New England Journal of Medicine* 325, no. 16: 1180–1183; Vincent Kiernan, "Diffusion of News about Research," *Science Communication* 25, no. 1 (September 2003): 3–13; Gunther Eysenbach, "Can Tweets Predict Citations? Metrics of Social Impact Based on Twitter and Correlation with Traditional Metrics of Scientific Impact," *Journal of Medical Internet Research* 13, no. 4 (2011): 123.

11. Daniel MacArthur, "On the Challenges of Conference Blogging," *Wired,* June 3, 2009, http://www.wired.com/wiredscience/2009/06/On-the-challenges-of-conference-blogging.

12. GrrlScientist, "Science Blogging for Scientists: Planting the Seed," *Living the Scientific Life (Scientist Interrupted), Science,* September 24, 2008, http://scienceblogs.com/grrlscientist/2008/09/24/science-blogging-planting-the.

13. Rosie Redfield, "Arsenic-Associate Bacteria (NASA's Claims)," RRResearch, December 4, 2010, http://rrresearch.fieldofscience.com/2010/12/arsenic-associated-bacteria-nasas.html; Carl Zimmer, "This Paper Should Not Have Been Published," *Slate,* December 7, 2010, http://www.slate.com/articles/health_and_science/science/2010/12/this_paper_should_not_have_been_published.html; Carl Zimmer, "Did Rosie Redfield Just Refute #areseniclife on Her Blog?," *The Loom, Discover,* August 2, 2011, http://blogs.discovermagazine.com/loom/2011/08/02/did-rosie-redfield-just-refute-arseniclife-on-her-blog/#.VN47iMZZUzV.

14. Paul Knoepfler, "My Year as a Stem-Cell Blogger," *Nature* 275 (July 2011): 425.

15. Quoted in Eva Amsen, "Who Benefits from Science Blogging?," *Hypothesis Journal* 4, no. 2 (September 2006): 10–14, http://www.hypothesisjournal.com/?p=665.

16. Drew Conway, "Ten Reasons Why Grad Students Should Blog," *Zero Intelligence Agents*, June 8, 2010, http://drewconway.com/zia/2013/3/27/ten-reasons-why-grad-students-should-blog.

17. Ed Yong, "Am I a Science Journalist?," *Not Exactly Rocket Science, Discover*, June 28, 2011, http://blogs.discovermagazine.com/notrocketscience/2011/06/28/am-i-a-science-journalist/#.VN6KucZZUzW.

18. From a video by Saunders posted in Peter Janiszewski, "Social Media for Scientists: A Lecture," Science of Blogging, October 12, 2002, http://scienceofblogging.com/social-media-for-scientists-a-lecture.

2

From Page to Pixel

A Personal History of Science Blogging

CARL ZIMMER

The Internet is relatively young, and science outreach on the Internet is even younger. Carl Zimmer, an award-winning author, journalist, and blogger at National Geographic, *discusses the history of the blogosphere, implications for the future, and his own transition from traditional journalism to becoming one of the world's best-known science bloggers.*

Today blogging is one of the standard ways in which we tell the stories about science. This state of affairs is relatively new. For those of us who entered the science-writing world back in the twentieth century—as opposed to the twenty-first—the memories of a life before science blogging are still fairly fresh. Understanding the origins of science blogging can help us do it better now, and to push it into fruitful new experiments.

My own memories of life before science blogging start around 1990, when I got my first job in the journalism business—as an assistant copy editor at *Discover.* At the time, it was one of the biggest of the many magazines focused on science. It had a robust circulation of over a million readers. And it had no connection to the Internet whatsoever.

In that format, science writing had a simple one-way flow from writer to reader. A writer would research a story and write it. An editor would edit it, a fact-checker would make sure it was accurate, a designer would lay it out in an upcoming issue, a printer would produce millions of copies of the magazine, and truck drivers and ship captains would deliver it to the world.

In this one-way arrangement, it was rare for us writers to hear from our readers. Sometimes someone would sit down with pen and paper and write out a letter to the editor. But we had little sense of our audience. We had no way of knowing how many people read a given story, or how many of them talked about it with their friends.

The technology that would turn our journalistic world upside down already existed at the time. Though journalists had little idea that the Internet even existed, scientists had been using it since the 1970s. I stumbled across the Internet in 1994, when I was interviewing a scientist about his work on simulations of black holes. He explained to me that I could see his simulations on my own computer—and he wouldn't have to send me a CD-ROM. I loaded Mosaic software onto my computer, and it carried me, to my astonishment, to the scientist's web page. It was as if I had been hurled from New York and landed in a chair next to him in his office in Urbana, Illinois, thousands of miles away.

Even as I came to appreciate the web, it would have been hard to imagine then that my own stories would someday jump into the screen, that most people would read my work online rather than in print. The modems were too slow, the computer memories too infantile, the monitors too pixelated.

As a science writer, my own transition to the Internet was motivated by practicality. In 1999, I left *Discover* to become a full-time writer of books and articles. I wanted a place online where I could display my magazine articles in order to persuade editors that I could write for them. I also wanted to post information about books I had written and links to places where people could buy them. I discovered that no one had yet claimed carlzimmer.com and started to build a website. The site was useful, but it was also a lot of work. The primitive software of the day meant I ran a huge risk each time I wanted to make the slightest change. It was like replacing a jet engine at thirty thousand feet.

I was therefore amazed to discover that a few people had websites that they updated *every day*—and some of them were writing about science. The earliest of those science writers I can recall include Chris Mooney (writing on science and politics), Razib Khan (human genetics), P. Z. Myers (evolution and development), and Derek Lowe (drug development). Their topics and politics varied enormously, but they all shared the same lively, personal style.

The medium they used also gave them a power that print could not offer. As soon as something happened in the news, they could write a piece of commentary and post it within hours—or even minutes. Publishing was as simple as pressing a key. The bloggers, as they called themselves, could incorporate photographs easily into their text. To back up what they said, they could link to original sources. And they offered readers an opportunity to respond, by providing comment threads.

Intrigued, I started playing around with blog software. I was attracted to blogging because I wanted to write about things that weren't very welcome in print publications, and I wanted to write in ways that didn't fit their style. Because I was my own publisher, I didn't have to ask anyone's permission to write what I wanted. In 2003, I launched my blog, which I dubbed *The Loom* (an obscure reference to a line in chapter 93 of *Moby Dick*). It's been an intimate part of my writing life ever since.

In hindsight, I can see that my experience was just a small part of a turbulent chapter in the history of journalism. Print publishing was beginning to slide. In the 1990s, magazines and newspapers were so lucrative that corporations gobbled them up. Debts soared on the assumption that the good times would never end, and that print would always reign supreme. The *New York Times* spent over a billion dollars buying the *Boston Globe,* reportedly because they had the best color printing presses in the country. Color printing, not the Internet, was the future of journalism.

And then the crash came.

Corporations tried to pay off their debts by squeezing bigger profits out of their publications. When the profits weren't forthcoming, they cut costs by slashing staffs. Special science sections vanished from newspapers; science writers were laid off. Editors became anxious about stories that wouldn't grab as many people as possible. No essays about altruistic slime molds, please.

That editorial fretting didn't stop newspapers and magazines from losing huge numbers of readers, many of whom shifted to the web. Meanwhile, the advertising that had buoyed magazines and newspapers began to evaporate. Classified ads migrated to Craig's List. Luxury ads also moved online. Sadly, most print publications didn't give serious thought to a better way to cope with the changes in journalism: by investing in good websites. For years, their websites were little more than copy-paste dumping grounds for their print edition.

Like other science writers, I did my best to tread water. I wrote freelance articles for magazines and newspapers, figuring out the sorts of stories that worked for each outlet. When I needed to write for myself, and for like-minded readers, I blogged.

The greatest pleasure I got from blogging was surprise. I would delve into strange corners of biology—a wasp that turns a cockroach into a zombie in which it can lay its eggs, for example. And I could see that people really did like to read about such stuff—and share it with their friends. The analytics for my blog showed me that I had readers from all over the world. I could see how other bloggers linked approvingly to the zombie post. Eventually the wasp ended up as a villain in a video game. A band posted a video on YouTube in which the members sang about the wasp's attack as a metaphor for a romance gone especially bad. I could see the unpredictable ways in which the things I wrote spread through the maze of culture.

Blogging also let me jump right into the biggest science news stories of the day. In 2005, a judge in Pennsylvania was hearing a case brought by parents complaining that creationism—in its latest form, "intelligent design"—was being slipped into their local school. Judge John Jones delivered a devastating rejection of intelligent design and his decision was posted online. I grabbed a copy and read through it, blogging as I read. As I updated my post, readers were having their own discussion in the comment thread, making collective sense of this historic moment.

I sometimes responded to creationists on my own blog. Traditional publications didn't see such responses as part of their mission. I disagreed, and used my blog to explain why creationist claims were wrong. By the time I had finished explaining how scientists know that the world is not just six thousand years old, I had explained geochronology—real science.

For the first few years of my experiments with blogging, some of my more distinguished colleagues in science journalism were baffled that I was "wasting" so much of my professional time. I was frustrated sometimes trying to explain why I enjoyed it so much. I couldn't get them to see the possibilities that blogging—both the software and the cultural practice—opened up for science writing. They joked about how I was going to end up living in the basement of my mother's house, blogging in my pajamas.

There's a hostility laced into such jokes. Many journalists saw themselves as professional gatekeepers, who used careful judgment to decide what kinds of science should become part of the public record, and to decide how their stories should be told. Now anyone could launch a blog and make a mess of things.

Professional journalists didn't just view bloggers as degrading the craft. They also viewed bloggers as an existential threat. By the mid-2000s, traditional science journalism was in a dire state. The pay that writers could get for their journalism fell. Instead of painstakingly researched investigations, editors seemed to favor superficial, quick blurbs—and lots of them. In addition to their print editions, these editors were now trying to fill their websites with what they now referred to as "content."

Somehow, the bloggers must be to blame. They had flooded the market with reading material—material they had produced not for money, but for the sheer pleasure of blogging. They undermined the work of real science journalists, and the whole edifice collapsed.

The idea that some pajama-clad basement-lurkers could destroy a major sector of the media is absurd. The real reasons for the collapse of traditional science journalism are more complex, and they stretched back long before the rise of science blogging in the early 2000s.

Today, things have changed far beyond what I could have imagined when I started out in journalism. From 1950 to 2000, American newspapers tripled their revenue from advertising, to $48 billion a year. Since then, revenue has crashed to $22 billion—a level not seen since 1950. Today there are fewer people employed by newspapers than in 1947.

Science reporting has been utterly transformed by this industry-wide change. During the 1970s and 1980s, U.S. newspapers set up new science sections at a steady clip until they reached a peak of ninety-five. Since 2000 most of those sections are gone—only seventeen remain. Many prominent science magazines, like *Omni* and *Science 80*, shut down.

These statistics are an ugly reality for people whose mortgage depends on the economics of journalism. But they are also a distraction from the mission of science journalism. We should judge the success of science journalism not by how many people it employs, but by how well it supplies readers with the stories of science. And by that standard, it is a huge success.

Traditional media have finally taken the Internet seriously. They see their websites as the core of their operation, where they can deliver news quickly and efficiently. Readers can now sit down with a tablet and read about science in newspapers and magazines around the world, from the *Guardian* in England to the *Jakarta Post* in Indonesia. A quick Google search can deliver even pre-Internet articles from digital archives.

Newspapers and magazines have stopped looking at blogs as the enemy and have started seeing them as an opportunity. They now realize that they can use the format to report quickly, to give their writers a more personal presence, or to build a community of readers through forums and comment threads.

Science blogging, I would argue, has become so mainstream that the term is becoming obsolete. As I write this chapter in 2014, there's an ongoing boom of new, innovative news operations—places like *Vox, Fivethirtyeight, Matter,* and *Mosaic*—that put science at the center of what they publish. These publications are purely digital and they use innovative ways to display information (interactive maps, for example), while hosting writers who don't have to hide their voices or their obsessions. They follow the tradition of blogging without feeling the need to use the word.

Those who are starting out in science writing would do well to understand this history. Some scientists set up blogs to emulate their scientist-writer heroes. They may envision themselves as the next Stephen Jay Gould or Lewis Thomas, for example. But a scientist writing essays in 2015 is doing something fundamentally different from a scientist writing essays in 1975. Scientists like Gould and Thomas could take advantage of the one-way, bottlenecked flow of information, publishing their pieces in, respectively, *Natural History* and *New England Journal of Medicine.* Today a blog post will not march off and find its own audience, because the structure of publishing has changed so much in the past few decades.

Bloggers today may not have the special platforms that Gould or Thomas had. But they have many, many consolations provided by digital publishing. Most important, they can use their new tools to bring innovative, meaningful writing about science to desktops around the world.

CARL ZIMMER is an award-winning freelance writer. He has written more than a dozen books about science including *Evolution: The Triumph of an Idea* and *Parasite Rex,* and has written for many outlets, including *National Geographic, Wired,* and the *Atlantic.* He writes a weekly column for the *New York Times,* and his blog is hosted by *National Geographic.* He also lectures at Yale University.

Carl is based in Connecticut. Find him on his website at http://carlzimmer.com or follow him on Twitter, @carlzimmer.

3
How to Set Up a Science Blog

KHALIL A. CASSIMALLY

If you've decided to start a blog, there are several ways to get started and options that are available to you. Khalil A. Cassimally, community coordinator at The Conversation UK, *goes step by step through the basics.*

When I started blogging around ten years ago, a blog, to many people, was the evolution of the leather-bound diary. Instead of chronicling the day by writing in pen on paper pages that no one would read, bloggers typed into an Internet browser window and clicked "publish." I did this too.

Now it's popular to blog about more than your life. The universe and everything are also frequent blog fodder, and setting up a blog is simple, rapid, and free. To create a blog, you merely have to choose a blogging service, come up with a good name, and have some ideas or views you want to share.

There are a number of blogging services to choose from. At the time of this book's publication, two of the most popular are Blogger.com and WordPress.com. Blogger was created in 1999 and WordPress in 2003; both provide users with all the basic blogging tools necessary to write and publish.

Blogger is owned by Google, so if you live in the Google online ecosystem (which you do if you use Gmail, Google Drive, Chrome, and/or Android), you may find it advantageous to host your blog there. You'll find yourself instantly at ease with Google's familiar clean and minimalist style. Blogger also allows you to change easily the appearance of your blog.[1] There are multiple templates to choose from, and the possibility of fiddling with attributes such as color and layout. You can even add Google services for your readers such as translating your posts with Google Translate, emailing your work to others via Gmail, and adding a "+1 button" to track reader interest. Blogger also comes with a built-in statistics page. This allows you to see how many people are looking at your blog generally and at specific blog posts, and shows you what other sites have led them to yours.

While Blogger is a simple, reliable blogging service, it may seem a little limiting to those who want to have more control over their blog's look and capabilities. For example, you might want readers to be able to access your posts based on when they were published. Or you may prefer to have links to some of your best posts. You can easily choose between these options and more with WordPress. Don't forget that you can also add multiple pages to your blog—think of an "about" page, a "contact" page, a "portfolio" page—to turn your blog into a proper website.

In addition to offering near complete control of your blog's look,

WordPress has a decent spam filter (which, trust me, is a wonderful defense against a barrage of Viagra ads). It also has good built-in social media integration, allowing readers to share your blog posts on Twitter, Facebook, and numerous other networks.

One hugely important benefit of WordPress is that it makes it easy to back up and send posts to other websites.[2] This allows WordPress to serve as what's known as a "content management service," or CMS, used to run entire websites.[3] Websites such as *Scientific American, Quartz,* and *Re/code* operate entirely on WordPress, so you can imagine the flexibility that it provides. This is something to consider if you intend to spin off your blog into a mini-media empire. Considering the myriad of possibilities provided by WordPress, it is perhaps inevitable that it is not as user-friendly as Blogger, but the added benefits are considerable.

Founded in 2007, Tumblr.com is one of the few successful social networks built around blogging. Tumblr is worth considering especially if you intend to post graphics with minimal text. As opposed to Blogger and WordPress, where each blog post is essentially a webpage, a post on Tumblr is a shareable entity that will travel around among Tumblr users. The uniqueness of the Tumblr experience started with a focus on the "share" button. Users can also follow you and get your updates the next time they log in. Because the emphasis on sharing is so obvious on this service, shorter, snappier, and media-full posts tend to receive more attention because they are quicker to consume and therefore shared more frequently.

In 2012 Ev Williams, the man who started Blogger and Twitter, created another blog outlet, *Medium*. I think writing on *Medium* is a beautiful experience. The blog editor is refreshingly simple, with the focus very clearly on your words and photos. It uses a WYSIWYG (what you see is what you get) editor. Like Tumblr, *Medium* is very network focused. You can add a post on *Medium* to a "collection," which includes posts written by other *Medium* users based on a certain topic. Users can follow collections, which help them discover your work. Your page on *Medium* shows just a few links to your *Medium* posts (for an example, see my page at *Medium*.com/@NotScientific). This emphasizes *Medium*'s focus on individual posts rather than an individual blogger.

To set up a blog on any of these services, you first need to create an account. The process is as simple as creating a profile on social media networks. Generally social media networks require a username, password, and a name for your blog. Your chosen username can help brand and identify you, and thus you should think about how you want to present yourself to the world. Do you want to tie your blog or social media to your real name? Or do you want a pseudonym in order to separate your writing from your personal life or other professional career?

Your blog's name is also worth some careful consideration. It is one of the first things that a new reader will see, and first impressions are important. All visitors have the ability to leave and never come back or never to come in the first place. When I discuss blog names with potential bloggers, I always ask them to choose one that has a connection with the themes they will explore on their blog. Once you think you have an adequate name, make sure you Google it to verify that no one else had your same eureka moment.

With your blog created and named, you are free to take over the world. You just need to find readers. My first few blog posts had zero comments: a disheartening experience shared by most new bloggers. Then, slowly, some of my blog posts began getting one or two comments. And gradually, ten or more comments. To find and retain readers, you need to tap into the particular communities of bloggers and readers who will find your blog posts especially interesting. This does not happen overnight.

Start with the Content

First, always strive to publish quality blog posts. Writing quality posts is time-consuming and draining, but stick to it. Publishing a blog post that you are proud of is rewarding and chances are that readers will enjoy it too.

Try to blog regularly. No one likes a blog that has not been updated in months. What if you run out of ideas? I have found that talking to friends, lab partners, or my parents has helped me to clarify ideas for posts and spark my enthusiasm. Another great way to keep myself blogging regularly even when I am not particularly full of ideas is to send out a lot of short (just three- to four-hundred-word) "quick-fire" commentary blog posts that capture my thoughts about a particular article or blog post that I had read recently.

When I write longer blog posts, though, I find it a good idea to jot down the structure of the posts before I start writing. I typically summarize each paragraph into one sentence and think about the link between the paragraphs. This helps to ensure that I don't get stuck halfway through the piece, not knowing how to proceed. I'm also a big fan of Hemingway's "dictum to end each day's writing in the middle of a passage, or even a sentence," as science writer David Dobbs eloquently put it.[1] When I start writing the following day, I then know what I'm going to write, so very rapidly get in "the zone."

Once you get in the habit of blogging, you will find it easier to keep going. The creativity floodgates of your mind are open and ideas for future posts will randomly but frequently pop up. Keep a note of those ideas so that you can explore them later on. Your developing audience will also motivate you to write with their constructive comments and indicators of approval, such as Facebook likes, Google +1s, and Twitter retweets.

Again, make sure you read other science blogs that cover themes similar to your own blog as well as those which you simply find interesting. In addition to a web search, you might try exploring the science blog repository *ScienceSeeker (http://scienceseeker.org/index)*. Reading other science blogs allows you to learn from other bloggers who have been at it for a long time. Study their techniques and let them inspire your own unique writing.

Index and Share

The next step is to get your blog out there. Shout! The biggest difference between a diary and a blog is that a blog resides in an open web. But this poses a problem: how do readers find your blog among the millions of others?

There are a number of ways to get your blog out there. Start by submitting it to search engines such as Google and Bing.[5] These search engines will index your blog and display links to it in people's search results. Having your blog appear on search results is one thing but having it appear among the top results will bring more people to your content. For this reason, it is a good idea to think about what's called search engine optimization (SEO), or the use of writing and coding techniques or strategies to lift a blog to a higher rank when people search for it.[6] A web search will bring up many articles about SEO, but Google itself offers some guidelines.[7]

You should also add your science blog to a blog repository. If you blog about peer-reviewed research, *ScienceSeeker* allows you to mark those specific blog posts with a few lines of code (which merely involves copying a few lines and pasting them into your blog post) that it will then identify in order to list your blog post among other such "research blog posts." In essence, *ScienceSeeker* gives you the opportunity to have your blog posts listed with those of others, including more experienced bloggers, and to be linked into the science blogging community.

Join the Community

A few years after I began blogging, social networks started garnering attention. Many of my friends were on a social network called hi5, which was like a stripped-down version of today's Facebook. After much reluctance, I caved in and created a hi5 account, which I used to rant about football, friends, and my teenage life. All in all, it was nothing that would be of much interest to anyone . . . except friends. My friends began to read my posts religiously, and conversations about our lives and experiences sparkled. I was engaging with a very real, albeit small, community.

I began writing about science around then and a few years later I joined a budding network of bloggers. The network already had a good, established community, and my science contributions tapped into it. By periodically sharing links to my science posts on my hi5 account, I slowly broadened my community, which led to more readers and more discussions.

The science blogosphere is close-knit. If you contribute good blog posts and comments, you will likely get more attention. So it is essential to engage. The main approach is through social media (Facebook, Twitter, and other social networks) and, of course, through blogging. Your aim should be to get your name out there. If people know you, they will click on your links.

A good way to find strong science bloggers is to visit science blogging networks (you might try first http://blogs.scientificamerican.com, http://blogs.discovermagazine.com, http://wired.com/category/science-blogs, and http://phenomena.nationalgeographic.com). Read the blogs that interest you and interact with bloggers on social media and in the comments. Expand your network.

Blog comment sections in particular are great places to interact. Take the time to compose well-structured and opinionated comments. If you feel that a comment is not enough to get your message across, write your response on your own blog and share a link in the comments for others to follow. Not surprisingly, comment threads are a great place to discover new bloggers. It stands to reason that if someone wrote a thoughtful comment, that person's blog might be worth reading as well.

In addition to interacting with other bloggers and sharing their posts, you should of course share your own content. For example, you may consider creating dedicated social media pages for your blog, for example on Twitter and Facebook. As you engage with people, they will start to notice you and follow you back. They may also begin to retweet your tweets to their followers or help to promote your work. Slowly, you will build your own base of followers who will click on the links you tweet.

The key to attracting more people to your blog is to make sure that visitors see that you have a unique voice and offer them great posts. Science blogging is sharing what you are passionate about with others, giving them the same buzz you get when you read about space, Darwin, or panda sex. It is about sharing knowledge . . . while hopefully getting some recognition. That's cool, too.

KHALIL A. CASSIMALLY is the community coordinator of *The Conversation UK* and has blogged regularly for *Scientific American, Nature,* and others. Previously he worked on science blogging networks for Nature Publishing Group.

Khalil is based in Mauritius. You can find him on his Facebook page, facebook.com/notscientific or follow him on Twitter, @notscientific.

Notes

1. "Blogger Template Design," Google, accessed February 13, 2015, http://www.chicagomanualofstyle.org/16/ch14/ch14_sec245.html.

2. “Moving a Blog,” WordPress, accessed February 13, 2015, http://en.support.wordpress.com/moving-a-blog.

3. “Showcase,” WordPress, accessed February, 13, 2015, https://wordpress.org/showcase.

4. Bobbie Johnson, “David Dobbs: ‘Exquisite Wisdom Can Be Hard to Follow,’” *Medium,* June 20, 2013, https://medium.com/@bobbie/david-dobbs-exquisite-wisdom-can-be-hard-to-follow-1a7f0644ce92.

5. “Website Owner,” Google, accessed February 13, 2015, http://www.google.com/submityourcontent/website-owner/; “Submit Your Site to Bing,” Bing, accessed February 13, 2015, http://www.bing.com/toolbox/submit-site-url.

6. “Search Engine Optimization Starter Guide,” Google, accessed February 13, 2015, https://static.googleusercontent.com/external_content/untrusted_dlcp/www.google.com/en/us/webmasters/docs/search-engine-optimization-starter-guide.pdf.

7. “Do You Need an SEO?,” Google, accessed February 13, 2015, http://support.google.com/webmasters/bin/answer.py?hl=en&answer=35291.

4

Using Science Art and Imagery in a Blog

GLENDON MELLOW

Photos and art can bring a simple written piece to life. The best science communicators will work hard to ensure that they use artists' work responsibly and with permissions. Glendon Mellow, science artist and blogger at Scientific American Blog Network, *can tell you how to make sure that both writer and artist benefit.*

The painting *Science Chess* was borne out of a tweet, proving that complex communication extends beyond the realm of prose. Nuanced ideas can be found in fewer than 140 characters—or even no characters at all. Human beings are visual, pattern-seeking, symbol-speaking animals, and our eyes can be captured in countless ways. An arresting image is an essential, enticing lede to a story,

"I'm thinking scientific accommodation of religion is akin to letting someone take your King's Rook off the board because you're winning."—*Science Chess* © Glendon Mellow, oil on canvas paper, 2008.

and it's important to consider the potential impact such images may have.

In the case of *Science Chess*, a tweet led to the painting, which led to a blog post and a contest to identify the symbols in the painting. (The winner received a print of the image.) This series of events made me wonder which symbols were easiest for scientifically literate readers to identify, and why some were almost impossible. I wrote all of those thoughts out on my blog and a discussion ensued, one that emphasized how images can help shape our understanding of scientific ideas, with or without text.

Consider, for example, how:

- The viscera-free medical illustration of the respiratory system in a doctor's office clarifies the complicated insides of our bodies
- The deceptive graphs used by climate-change denialists highlight brief cooling periods instead of the overpowering evidence toward warming
- The dominant feature of print magazine covers is the central image, not the stories within or even the magazine's logo
- Feathered dinosaurs and exoplanets cannot be photographed, yet we still believe we know what they look like
- Mind-bending visual exercises are used effectively to depict the complicated world of quantum physics

Images Are More Than a Frill

Visual communication has been around for thousands of years. We are visual beings. There's a reason that in video games high-end graphics have won out over text-based adventures. It's the same reason that apps consist of a symbol and splash of color instead of just text.

The Internet is the same. With faster and faster processing and delivery speeds, larger and crisper images are becoming the norm. Graphics I first optimized for speed and quality seven years ago on my SciArt blog now look terrible on my iPad. This trend is affecting all major social media sites.

As I write this, giant banners on Twitter are about a month old. Google+ and Flickr both support huge amounts of space to store hi-res images. Blog templates everywhere are copying the screen-wide images of Medium and SquareSpace.

We do not put up big images just because we can.

A June 2013 *MIT News* post reported a study by Stephanie Hatch that ranked posts on the MIT Facebook page.[1] She found that among the top twenty posts in a month in terms of user engagement, 70 percent had images. According to social media scientist Dan Zarella, inline images added to Twitter in 2013 were 94 percent more likely to get retweeted than tweets with a link to an Instagram image that you had to click to see.[2] Images aren't just visuals, they actually increase the reach of your entire message.

The implication for blogs is that images can go a long way toward guiding people to your post and making sure it gets shared. They are not just pretty packaging; they are an effective means of communication that no blogger can afford to do without.

Think about the Image You Want

You've crafted a science blog post you're proud about. What's next?

Don't just jazz it up with a random, tenuously related stock photo. Think about the type of image you want. Are you worried that the topic might seem boring at first glance, and readers may need to be enticed to stay with it until the end? Try fine art or concept art, both of which can be immediately intriguing and work well with editorial pieces. Need to reassure people that the thing they think is scary (sharks, vaccines, GMO-food) isn't? Bust out cartoons and bright colors. Is your message serious? Skip the cartoons and go with some simple infographics, something meme-worthy to spread the message around.

Don't undermine your own message. A post on the efficacy and importance of vaccines doesn't gain anything with an image of a child crying while getting stabbed with a cold blue needle. The happy child talking to a doctor whose face we can actually see will be more effective at enhancing your message. You might think it's funny to show a photoshopped square tomato, but to persuade people nervous about GMO-foods, consider wholesome, down-on-the-farm photos or that still life with a bowl of fruit from your art history class.

The SciArt Movement

What if you can't think of the perfect image? Where do you turn? There is a whole world of scientifically literate artists out there whose work is often described under the umbrella term "SciArt." Find some science artists, and seek their advice.

Scientific literacy has been evident in the fine arts for a very long time. The Cubism of Picasso was arguably informed by the physics of a fourth spatial dimension. Seurat played with optics and a type of pixelation. The *Garden of Earthly Delights* by Hieronymus Bosch depicts fountains made from an alchemical distillation apparatus.

SciArt as a movement is growing, quickly and with complexity. Artists using traditional materials are increasingly choosing to depict science as a subject. Others are using the tools of science to create art. (Some in the subfield of bioart, for example, have used luminescent bacteria for screen printing.) What's more, these artists are connected as never before, mainly through blogs and other social media tools, which have added visibility and volume to this emerging art form. Science matters, and SciArt will not let you ignore it.

What Is SciArt?

First, a necessary digression about the taxonomy of that tricky word "art." We're talking about visual art, and even within those discussions there is a common misuse of the term "art." If Art is the Kingdom, and Visual Art the Phylum, Fine Art is a Class. And yet often when people discuss "art" they mix up the large term Art with the smaller term Fine Art.

From Fine Art paintings to bioart, from medical illustration to paleontography, from cartoons to data visualization, scientifically literate visuals have never been more plentiful, varied, and easy to find than they are today.

For a tweet-length definition, SciArt is visual imagery that explains science, uses the tools of science, or demonstrates scientific literacy, often to provoke discussion.

Where to Find SciArt

This is not a comprehensive list, but here are some sites where you can find SciArt. Many of the artists are ready to discuss their work, and are easily approachable via Twitter, blog comments, or email. Perhaps you would like to use their work on your blog (with permission), or to commission something new. And who knows, perhaps your writing may be transformed by a partnership with an artist that lasts for many years. The partnership between writer Eric M. Johnson and illustrator Nathaniel Gold at *The Primate Diaries* is a fine example (http://blogs.scientificamerican.com/primate-diaries).

- Twitter hashtag: #sciart
- Art & Science Journal: http://www.artandsciencejournal.com
- ART Evolved: http://blogevolved.blogspot.ca
- Art.Science.Gallery: http://www.artsciencegallery.com
- ArtPlantae Today: http://artplantaetoday.com
- Association of Medical Illustrators: http://ami.org
- The Finch & Pea: http://thefinchandpea.com
- Guild of Natural Science Illustrators: http://www.gnsi.org
- Mad Art Lab: http://madartlab.com
- Morbid Anatomy: http://morbidanatomy.blogspot.ca
- Phylo, the trading card game: http://phylogame.org
- PhyloPic: http://phylopic.org
- SciArt in America: http://www.sciartinamerica.com
- Science Artists RSS Feed: http://friendfeed.com/scienceartists
- ScienceArt Community on Google+: https://plus.google.com/communities/113301495673079979034
- Southern Ontario Nature & Science Illustrators: http://sonsi.ca
- Street Anatomy: http://streetanatomy.com
- Symbiartic: http://blogs.scientificamerican.com/symbiartic

The Not-So-Scary World of Copyright

Most bloggers initially hear about copyright when they hear that someone has violated it. Copyright for images varies a little country by country, and with people talking about open copyright, Creative Commons Licenses, GNU licenses, and cases of litigation, it can all seem overwhelming. But for a science writer, it's not so bad. It all comes down to citing sources.

If you see an image that would be just perfect for your blog post,

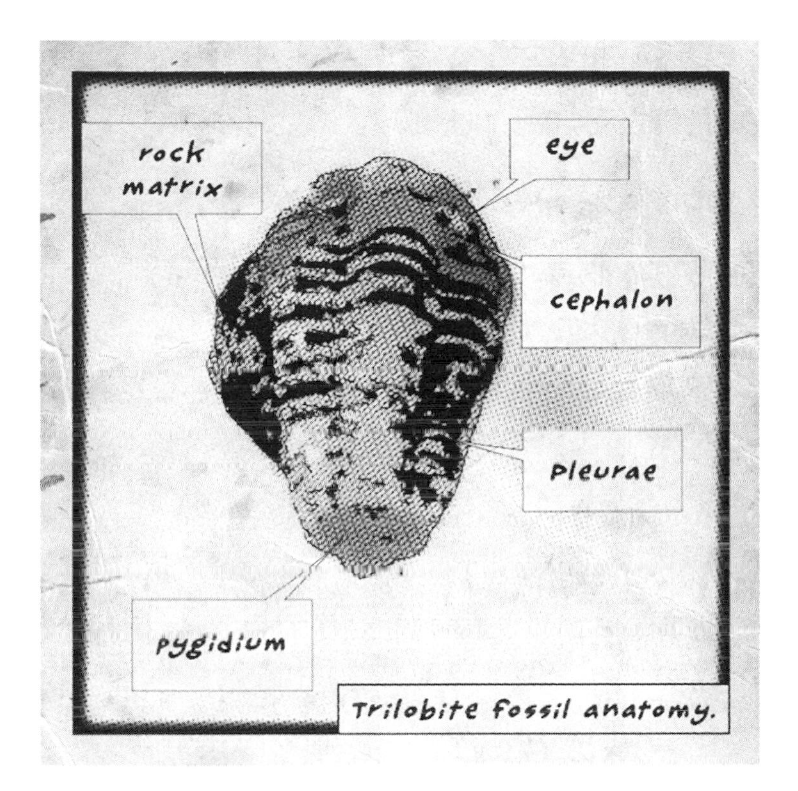

Why not make your own images? This one was made using an iPhone 4 with the Manga Camera app to simplify a photo, and the Halftone app to add captions. Image © Glendon Mellow.

remember that you can reach someone on the other side of the planet more easily than ever before. Make contact on Twitter or a blog comment or via their page on *DeviantArt*, an online social community for artists. Try email for a formal touch. Simply explain why you think the image is effective and how you would love to include it on your blog post for outreach or education. See what happens. Getting permission directly from the artist is the best and easiest way to use amazing images on your post.

The next step is to cite your source. You wouldn't dream of presenting a scientific study without referring to the paper, so make sure you credit the artist with "©Artist's Name." Link it back to that person's primary online portfolio, preferably in the caption below the image or somewhere else nearby.

Do not offer "exposure" to the artist as a form of payment. He or she is on the Internet and has the same chances for exposure that you do. Getting paid in exposure is frequently and cynically joked about by artists and illustrators. See @forexposure_txt on Twitter if you need to be humbled. But of course, many beginning bloggers don't have money. Just be up front about it. Try something like, "I do not have a budget on this project to offer you money, but I think it would fit our post perfectly. If we have your permission, what link would you like me to include on the attribution?" While pretty much all artists expect compensation for new work, many will happily allow their past work to be reused on a blog post.

Sometimes, despite your best efforts, you won't be able to reach the person behind the perfect image. In these cases, learn how to refine a Google Image Search. You can choose to search only works with a Creative Commons License (CCL), and many sites, such as Flickr and DeviantArt, will tell you right on the page whether the work is under CCL. There are a number of different Creative Commons licenses, so make sure you're abiding by the artist's terms. All Creative Commons terms require citing the artist as a source and suggest that doing so with a link is best.

If you find the right image via Wikipedia or Google Image Search, do not cite the source as "Image: Wikipedia" even with a link. Click through on Wikipedia or Google and look at who uploaded the image. Often that is who you should credit, though if it is a photo of, say, a painting by da Vinci located in the Prado, you should simply link back to the upload page on Wikipedia but write "Image by da Vinci, Prado Gallery." Clarity is the most important rule.

These guidelines are important to follow for reasons beyond simple responsibility. The tools that artists use to police their own work are growing (including Google Search by Image and Tineye), and, as noted above, the world of science artists is becoming more organized and tightly knit. I can think of a few examples over the past several years where a blogger used multiple pieces of an artist's work and cited only "Image via Google" (or even worse, cropped out the artist's signature or watermark). The result was swift justice. Artists will first ask nicely for the oversight to be corrected. If this has no effect, they will ask other artists to help flood comment pages or make complaints to ISP or site hosts until the errors are fixed. Science communication is a small field, and chances are the people most interested in your post will be able to recognize your sources, even if improperly credited.

The formula is simple: get permission, and cite the artist with his or her name and a meaningful link.

How to Share Your Own Images Online

This chapter was written primarily for science communicators who do not necessarily identify themselves as artists. But there is a growing and rich world of science bloggers who communicate visually. Here are a few best practices for using your own images online in the service of science communication:

- Watermarks—use your URL, preferably if it has your name in it, but don't make it huge or obscuring.
- Get comfortable with reverse image search tools, which will help you to keep an eye on any of your work that's going viral and has your name cropped off.
- Build your network. Your blogroll, sites you comment on, and whom you follow on Twitter should include potential

clients as well as other artists. Not for a hard sell, but to forge friendships and meaningful connections. The same connections can sometimes alert you to infringements on your work.

- Forget separating the art from the artist. Your voice identifies you as much as your artwork does.
- Let people share your work with attribution. Don't worry about re-posting older work. There is always someone new who hasn't seen it.
- Be professional. Use agreements when making new work and have a plan.

Imagery of all kinds has a place in science communication and on blogs. Be proud of what you have to offer.

GLENDON MELLOW is a fine artist and illustrator. His artwork is featured in science blogs, magazines, books, and exhibits. He also gives talks about social media, art promotion, and the growing field of SciArt. He writes the blog *The Flying Trilobite* and is part of *Symbiartic,* on the *Scientific American Blog Networks.*

Glendon is based in Toronto. Find him on his website, www.glendonmellow.com, or follow him on Twitter, @FlyingTrilobite.

Notes

1. Robyn Fizz, "The Big Picture: Using Images in Social Media," *MIT News,* June 11, 2013, http://newsoffice.mit.edu/2013/the-big-picture-using-images-in-social-media.

2. Dan Zarella, "Using Images to Get More Retweets," DanZarella.com, October 7, 2013, http://danzarrella.com/use-images-on-twitter-to-get-more-retweets.html.

5

Building an Audience for Your Blog

ED YONG

"If you build it, they will come" might apply to a baseball field, but probably not to your blog. If you want people to see your work, you will need to find and reach out to them in creative, thoughtful ways. Ed Yong, an award-winning blogger with National Geographic, *will teach you how to reach your intended audience.*

If you Google "top tips for blogging," chances are you'll get a list of sites that tell you similar things: write short, punchy posts, and the shorter the better; try and post every day; be personal; and so on. Feel free to ignore these tips. The great joy of blogs is their diversity. Some flit between topics; others have a sniper's focus. Some offer a nearly constant fire hose of updates; others offer monthly drips. Some are intensely personal; others subtly so. Some offer canapé-sized snippets; others go for meaty long-form fare. There are few consistent factors that make blogs successful or enjoyable, and assuming that such factors exist ascribes a special mystique to blogs that they don't deserve. A blog is just a container. It's a channel for communication, much like email, or radio, or talking. The rules for running a good blog are exactly the same as those for acing *any* form of communication: you have to have something worth writing about, and you have to write it well. Le fin.

Playing the Long Game

Blogging is a marathon, not a sprint. Unless you happen to launch a new blog under the umbrella of an established media brand, or you already have a substantial following online, you will probably spend a lot of time without many readers. When I started *Not Exactly Rocket Science* as an independent WordPress site, I had just a few hundred page views a day for at least eighteen months, or until I joined the *ScienceBlogs* network. That was a bit dispiriting, but it didn't matter. I wanted to write. I had an itch that needed to be scratched. And the fact that some people were reading—regardless of how few they were—was valuable, rewarding, motivating. Small audiences matter; they're a necessary stepping-stone toward big audiences. And it takes time—months, maybe years, of effort—to build a big audience. There's no way of shortcutting your path to greater traffic.

Actually, I lie. It is possible to cheat your way to page views. You could sacrifice quality for quantity, and post an incessant stream of YouTube videos, funny images, and clips from other people's work. With more posts, a small amount of traffic for each one translates to a bigger total. You could take a leaf out of the Buzzfeed and Upworthy book and go for "linkbait" headlines like "This One Freaky Fish Crawled Out of the Water 400 Million Years Ago; You Won't Believe What Happened Next" or "16 Genetic Diseases You Need To Know About."

But all of these tactics will probably seem annoyingly obvious to your readers and to other bloggers. So let me revise my earlier statement: there's no way of shortcutting your path to greater traffic, because page views aren't the same as an audience. Page views are a short-term reward and one that's easy to game. An audience is a long-term reward—a growing cadre of readers who know and follow your work because of its quality, not because of cheap tactics. The latter is immensely more valuable but takes longer to get.

In the meantime, you'll need to keep yourself motivated. First and foremost, have fun with it. Write for yourself, so that the act of crafting a post is rewarding in itself. You will need some form of self-contained feedback that's independent of page views, comments, and other external metrics to keep you going during those early months when you're finding your feet. For me, it was the process of writing. Coming up with a good metaphor, finding a sharp intro, and breaking something complex down into accessible morsels were rewarding acts in themselves. They kept me going until other rewards manifested.

Remember: blogging is a marathon, not a sprint. Also remember: the more you write, the better you will get at it. New writers who suddenly join big networks often complain of stage fright. There's a lot to be said for quietly getting better, faster, and more confident at putting your thoughts into words so that when the limelight falls on you, you are ready to perform. If you worry that you're not good enough, or that no one would want to read what you write, bear in mind that everyone thinks that at first. But it's also true that you may well be right about that, and if so, you'll continue being right until you get enough practice in.

Finding Your Niche

The world of science blogging is a crowded one and it can be hard to make yourself stand out. First, a caveat: you may not want to stand out. It is fine to blog whenever you feel like it as a chance to scratch a writing itch, and with no expectations of reaching wide readerships. If that is the case, you may want to consider writing occasionally for sites that take guest contributions, rather than starting your own blog. *Scientific American*'s *Guest Blog, The Conversation,* and *Medium* are all good options.

For those who are starting their own blog, let your curiosity and your passions guide you. Blogging isn't a job. You won't get assignments from a boss. The drive to write has to come from you, and it will come more easily if you write about things that you love.

Some bloggers, myself included, cover a wide range of different topics. I was, and still am, led by my curiosity. I write about things that interest me and let my own visceral reactions guide my choice of topics, with the constraint that I avoid topics where my foundations are weak, like physics. Others stick to specific topics and have become indispensable resources for people interested in the same—Maryn McKenna covers germs and food production on her *Superbug* blog, while Vaughan Bell and Tom Stafford at *Mind Hacks* cover all things brain-related. Neither a narrow focus nor a broad one is inherently better. Going narrow might help you to focus on what you want to write about, and to find stories that no one else is doing; it might also hamstring you. Going broad gives you more material to draw from; you also risk drowning in so many possibilities that you get stage fright, or getting burned if you write about something way outside your expertise without doing your due diligence. Some people thrive in total freedom; others like constraints. Choose your path based on your temperament.

You could focus on explaining new research. My fellow Phenomena bloggers Carl Zimmer, Nadia Drake, Virginia Hughes, and Brian Switek are masters of this form. Or, like the collected bloggers of the *Scientopia* network, you could talk about issues in academia and life as a scientist. If you're a scientist, you could blog about your own research—John Hutchinson's *What's in John's Freezer* is a favorite. If writing critiques is your thing, you could specialize in busting pseudoscience, like Ben Goldacre does in his blog *Bad Science.*

The type of posts you write can vary too. Some people focus on creating their own compelling posts. Others also point their readers to everyone else's compelling posts. Sites like *io9* and *Boing Boing* do this exceptionally well, by redirecting their own readers to interesting material curated from across the Internet. People visit these sites for their own excellent content, but also because they trust the authors on these sites to show them where to find more excellent content. This is why, on my own blog, every Saturday I compile a list of links to good reads.

Regularity is important, although because many readers find blogs through social media and aggregators, perhaps not as important as it used to be. There are bloggers who write daily, weekly, or monthly. Some write for multiple outlets, and use their blogs as a roundup, or to place the pieces for which they can't find a home elsewhere. Really, the only bad move is to not write at all. Months of silence aren't ideal for building an audience. Aatish Bhatia, now at *Wired,* turned *Empirical Zeal* into a well-regarded science blog by writing consistently eye-opening explainers that got people talking, but doing so only once a month.

Some people go it alone. Others find like-minded colleagues to start a group blog like *Deep Sea News* (marine biologists) and *Last Word on Nothing* (journalists). The group strategy can provide safety and strength in numbers, if you don't have the energy to blog often enough on your own.

Regardless of your predilections, what matters is carving out a niche. That could be an unusual topic that no one else is covering. Ivan Oransky very quickly turned *Embargo Watch* and *Retraction Watch* into big hits by focusing on two areas that no one else had their eye on. Alternatively, your niche could be a knack for deep critical analysis; humor; illustrations or videos; an insider's perspective on a particular field; or just wondrous storytelling acumen.

Find something that makes you stand out. If you write yet another collection of short posts about new science, you are competing against the hundreds of news sites that already exist. If you write a screed about homeopathy or creationism, you are competing against the hundreds of such screeds that have already been published.

Remember that readers will keep coming back not just for what you say, but how you say it. They are not just interested in the things you write about, but in you as a writer. Your voice and your personality are what form the essence of your blog, so do not be afraid to let those show. Don't feel the need to define these from the start. You don't need a mission statement to blog; you need only an Internet connection.

Thinking about Readers

The classic mistake that people make when they start blogging is to casually type away as if anointing the world with their Very Important Thoughts, rather than to consider the people for whom they are writing. Who are they? What do they care about? How much do they already know or understand? Heed the words of famed British science writer Tim Radford:

> When you sit down to write, there is only one important person in your life. This is someone you will never meet, called a reader.

> You are not writing to impress the scientist you have just interviewed, nor the professor who got you through your degree, nor the editor who foolishly turned you down, or the rather dishy person you just met at a party and told you were a writer. Or even your mother. You are writing to impress someone hanging from a strap in the tube between Parson's Green and Putney, who will stop reading in a fifth of a second, given a chance. So the first sentence you write will be the most important sentence in your life, and so will the second, and the third. This is because, although you—an employee, an apostle or an apologist—may feel obliged to write, nobody has ever felt obliged to read.

Different bloggers target very different audiences. Some, like me, are aiming for intelligent but nonspecialist readers. Others are writing for kids. Still others are addressing scientists within a specific field. If you're not sure who your readers are, ask them—every year, I create an open thread on my blog where I invite readers to de-lurk and say something about themselves, their background, and their interests. Your choice of audience will determine the language you use in your writing, the level of detail you go into, and the choices you cover. This is something you should probably consider in the infancy of your blog. Picture your ideal readers in your head: who are they?

If you're writing for a layperson, technical jargon is immediately off-putting (and linking to a definition is a lazy, inefficient solution; most people will not click on that link). A good guide is to assume that your readers last encountered science when they were in high school, and to use terms and concepts that you yourself were familiar with at that point in your life. Everything else gets explained. That's how newspapers operate.

If you're a scientist writing for peers, you can probably get away with including some jargon or lofty assumptions about background knowledge. But even then, remember that doing so can alienate other scientists who aren't in your field (and who might benefit from your post). If you're a quantum physicist, your writing may sound like Dothraki to a geneticist, even if she's a professor. Pop culture references have a similar effect. When I wrote "Dothraki," to some of you I may as well have been speaking Romulan.

In my view, regardless of your imagined readership, it always pays to make your blog as accessible as possible. This is particularly true when you start, because you are heavily reliant on word of mouth and on people stumbling across your blog via social media, search engines, or incoming links. One of the most important ways of building an audience is a passive one: avoid alienating incoming readers.

Language will help. So will structure—the way you arrange your thoughts and ideas into sentences, paragraphs, and posts. You could use the standard inverted pyramid of news stories, where the key details appear at the top. You could use storytelling techniques to describe a quest, or a specific experiment, or a moment in a scientist's life. You could do a straight Q&A. Blogs give you all the freedom you need to find your own voice, without the constraints of word counts or editors. They allow you to experiment with different styles and structures and write in the way that feels most natural to you, without falling into the stylistic monoculture that plagues most of the mainstream media. (The blogs that I named in the previous section are all tremendously diverse in terms of their writing style.)

But the freedom of blogging also has its perils: it lets you get away with publishing bloated, meandering, self-indulgent, narcissistic posts that would never see ink in a magazine. Good blogging, then, is a perverse and fiddly act of balancing the freedom to express yourself with the restraint that will make you readable. It's about remembering that the rules of good writing are relevant whether you are writing for a layperson or a scientist, and that once you know those rules, you can work out how to break them effectively. It's about writing for you, while also writing for a reader. It's about exploiting the new niches created by this Cambrian explosion of science writing opportunities while acting as your own strict editor. But ultimately, there is no secret recipe for a successful blog. What works for one person may not work for another. The only constant—the only tip that always applies—is to have something to write, and to write it well.

ED YONG is an award-winning freelance science writer. His blog *Not Exactly Rocket Science* is hosted by *National Geographic*. His work has also appeared in many publications, including *Wired*, *Nature*, the *BBC*, *New Scientist*, the *Guardian*, the *Times*, *Aeon*, *Discover*, and *Scientific American*.

Ed is based in London. Find him on his website at http://edyong.flavors.me or follow him on Twitter, @edyong209.

6

Ethical Considerations for Science Bloggers

JANET D. STEMWEDEL

Scientists and science writers want to communicate about the latest scientific findings, the process of knowledge-building, and the experience of being a member of a scientific community. But ethical pitfalls can occur as bloggers develop relationships with readers and experts. Scientist, philosopher, and blogger Janet D. Stemwedel considers the responsibilities that should accompany those relationships, offering practical advice on ethics for those delivering scientific content online.

Blogging Brings Responsibilities

Science blogging is an activity that takes many forms. Sometimes it resembles journalism, sometimes punditry, sometimes pedagogy.

At times science blogging is a first-person account of engaging with a scientific finding or building a piece of scientific knowledge, or of being part of a scientific community more generally. But whatever the style of the blogger, writing a blog differs fundamentally from writing a notebook because it assumes an audience.

Blogging about science, therefore, puts the blogger in a relationship with readers—and so, like any situation that puts human beings in relationships with each other, involves ethics. I've spent almost a decade blogging about ethics in science (and longer than that teaching ethics as a philosophy professor at San José State University), so I've had occasion to think hard about some of the ethical dimensions of human interactions both within science and across the Internet.

Luckily, being an ethical science blogger doesn't require that you decide whether Kant or Mill or Mencius knows what *really* constitutes "the good." Rather it's a matter of thinking seriously about your responsibilities to your craft and to your audience and then conducting yourself in a way that balances those responsibilities with good writing.

All blogging creates relationships with readers. Blogging about science means you're also engaging (whether deeply or superficially) with a body of knowledge, a process for building that knowledge, and communities of practitioners engaged in that process. Arguably, this engagement imposes a kind of duty to be accountable to the world that this science is describing—to attend to the empirical facts and be honest about what is known for sure, what is probable, or what is possible. *Not* being accountable in this way means departing from the ethical commitments central to the scientific enterprise, a step that would put you dangerously close to the positions taken by purveyors of pseudo-science. Blogging about science may also make you responsible to scientific practitioners, since what you blog about can influence how the public understands what they do and who they are.

Before You "Publish": Questions for Ethical Reflection

Consider this before you step onto your virtual soapbox: your blogging (or tweeting, or viral video creation and distribution, or what have you) has the potential to influence lots of people, and that influence may not always be in line with your goals. Thinking about those relationships and those goals before you put a post online for the world to see can help you do better by both, so before you publish a post, pause to ask yourself:

- What am I hoping to accomplish with this post?
- Whom will this post help? Whom could it hurt? What's the likely balance of good and bad consequences if I publish the post this way?
- To whom do I have duties that are relevant to whether I publish the post this way? Are my duties such that I *must* publish it? That I should significantly change the post? That I shouldn't publish it at all?
- Would a good science blogger (someone I'd respect—someone I'd want to be) publish this? Or could publishing this post undermine my ability to blog about science in the future?

The next step is to look beyond your goals and intentions, to the responsibilities that you have to your audience and the world beyond. What kinds of duties could you possibly have to legions of people whom you may never meet in a non-virtual space? Two duties that should be uncontroversial—because they are ways we expect our fellow human beings should treat us—are be honest and be fair.[1] Particular situations and relationships can give rise to more precise obligations, but they're usually some variant of these two.

Be Responsible to Your Story

The type of story you're trying to get right in many ways determines what steps you will take to be responsible to your audience. If you're describing a scientific finding or theory, you still want to get the factual details right. Does this mean you have a duty to read the primary literature presenting the finding or theory? If you can access and read it, you probably should, since this is where the details are described.

You may find you need help with this step, however, especially when the journal articles you are reviewing are terse or jargon-laden. Maybe you can ask the researchers themselves for a more accessible explanation of what they did, or rehearse your explanation for them to see if it correctly captures what they found and how they found it. Possibly other researchers in the field could help you work through the journal article or give feedback on your post; even if it's not their research, they are part of the intended audience for the primary literature.

Getting the story right may also involve shedding light on the larger context of a scientific finding—the environment in which the research is done, the practical questions motivating seemingly esoteric research projects, what scientists already know in this area, and how the new finding builds on that understanding or challenges it. In addition, you should accurately convey the level of certainty or uncertainty that accompanies the finding.

Sometimes the subject of your story is a person within the scientific community. Getting the story right might then require giving your subject a say, whether within the story itself or as a rebuttal after the fact. It also means being careful not to make claims that go beyond your evidence—say, about their motives or their emotional states. There is controversy about whether it's appropriate to let your subject see your post and require changes before you publish it (which is something that traditional journalistic ethics frown upon), or whether you have a duty to seek comments from people you write about before you publish that writing on your blog. Your best course is to consider what, if any, kind of interaction with your subject will strike the optimal balance of fairness to the subject and honesty in the story. And if you're writing online in a professional capacity, you should check with your editor.

Be Responsible to Your Readers

Getting the story right, whatever that story is, is part of the honesty you owe your readers. You also owe it to them to be clear about *your* level of certainty in the claims you're making—what you know on the basis of evidence, what you believe based on a hunch, and which parts of the story you really don't have any confidence about. Don't convey the impression that your conclusions are more certain, better evidenced, or less ambiguous than they actually are. Link your sources so your readers can evaluate them on their own if they want to.

Tell your readers where you're coming from—your background, your expertise, and the goals you hope to accomplish by blogging. Don't use your expertise as a cudgel, but instead explain how it informs your take on things.

Expose your biases (at least the ones you know about), and ask for help in noticing where your biases may be driving the story you're telling. When you make mistakes, be transparent about admitting and fixing them. Editing your post without comment (for anything more significant than grammar, spelling, punctuation, or other typos) may give the appearance of erasing your mistakes rather than acknowledging them. Moreover, it can shift the focus from the story you're trying to get right to the subject of your credibility. Demonstrating a commitment to accuracy and fairness does a lot more for your credibility than defensiveness or deflection.

If blogging is writing that assumes an audience, commenting makes blogging a two-way mode of communication. This means you don't need to imagine whether you're succeeding in reaching that audience; you can find out by asking. Take advantage of the comments function to engage with your readers and find out where they're coming from. Comments can help you learn what they want to know, what questions they think are interesting or important, and what kinds of explanations reach them. Listen to this feedback and use it to make sure you're neither pitching your posts over their heads nor insulting their intelligence with overly simplistic explanations.

Be honest about the kinds of conversations you're willing to host on your blog and do your fair share to cultivate an environment that supports them. Some research suggests that hostile comment threads may influence readers' judgments about the *factual* content of the post that precedes the comments—so you should consider setting clear boundaries for how commenters may engage with you, and each other, in an explicit commenting policy (tips for creating such a policy are given in Chapter 24).[2] Then enforce that policy to provide your readers (even the ones who don't comment) the environment that you promised them.

Be Responsible to the Larger Community

If you're blogging about science—especially if you're also a scientist yourself—your blogging may influence people's understanding not just of science but also of scientists. Visibly striving to be accurate and precise, to convey context and uncertainty, and to correct mistakes when you make them is a way to display values that are supposed to be part of the scientific enterprise, too. By contrast, being dismissive of questions and concerns of nonexperts, acting defensive when you err, pulling rank, or being a jerk can feed a negative impression of scientists or science communicators. Behave the way you think a good blogger (or scientist, or employee of your organization, or human being more generally) should behave. When others in your blogging community or scientific community behave badly, communicate that you expect better.

Holding others in your community to high standards is complicated by the fact that science bloggers don't all have the same goals and interests. It's important, then, to recognize that it's not a moral failing if another blogger doesn't have the same agenda that you do—for example, if that person chooses to blog about personal experiences in science while you prefer to blog about peer-reviewed scientific research. If another blogger misrepresents a finding, leaves out crucial context, draws an unwarranted conclusion, or heaps abuse on a subject or the readers, it is fair to explain the problem and express your disapproval. If you simply don't like a blogging style, however, realize that this may be a matter of taste and let it go. There is no single "correct" way to be a science blogger (although there are many opportunities to mess up), and the science blogosphere undoubtedly serves more people when it contains a diverse assortment of blogs and bloggers.

Be part of that blogging ecosystem. Engage with other relevant conversations, recommend them to your readers, and cite them, especially when they've influenced your own thinking. A link costs you next to nothing.

When making your recommendations, try to become aware of your biases. Which blogospheric voices do you reflexively treat as authoritative? Which do you ignore, perhaps without noticing that you're doing so? It is surprisingly easy to create and sustain patterns of inclusion and exclusion in your blogging and in your engagement with others. Recognize that challenging your own biases will help you to boost enthusiasm for and engagement with science.

Blogs are part of the media landscape; in fact, recent court rulings assert that bloggers count as journalists.[3] This means that your conduct could be taken as reflective of the media as a whole. You may be sent press releases with embargoed information, or offered access to researchers or free books to review. Try to behave as you expect a responsible journalist would—or to clearly distinguish what you are doing from journalism to avoid confusion. Existing codes of journalistic ethics can be an important source of guidance.[4]

Be Responsible to Yourself

While you're noticing your duties to others and the potential impacts, positive and negative, of your blogging on your readers, don't forget that *your* needs and interests matter in your ethical calculations. Don't let what you owe your readers and the wider community overwhelm your ability to take appropriate care of yourself.

Make sustainable choices about how frequently you post and how much time you allow for researching a topic and fine-tuning the writing. Don't agree to deadlines (or set deadlines for yourself) that will require you to sacrifice accuracy or clarity. If you invite your readers into a conversation with you, think hard about how much time you can really devote to the conversation you've promised, and about how much emotional energy you can spend moderating inflammatory or infuriating comments. If you need some breaks to keep up your energy and enthusiasm, find a way to take them.

Whether you're blogging under your real name or a pseudonym, draw sensible boundaries about how much personal information you want to share. Readers may feel more connected to you if you write about your family background, your geographical region, or experiences connected to your race or gender or disability—but you can also decide that some personal details are private. As long as you're not misleading your readers, you get to choose which parts of yourself they are allowed to see.

Decide what kinds of risks you're willing to undertake. Blogging about some topics may expose you to campaigns from paid commenters from corporate lobbying groups, targeting by animal rights groups, or personal threats to your safety. Getting help from an unnamed source may expose you to legal action if you try to protect his or her anonymity. This doesn't mean you shouldn't write about these topics or get help from unnamed sources, but be aware of the risks and find support and guidance from others who face them.

Finally, unless your blog is just paid work for someone else, regularly revisit the question of why it is you blog. Are you getting what you want from the experience? If not, what can you change to bring the fun back? Especially in the blogosphere, there is no reason that being ethical shouldn't also be fun.

Ethics, fundamentally, is a matter of sharing a world. Being an ethical science blogger (or videographer, or tweeter) means sharing that world not only with scientists who are trying to build an accurate picture of that world's features, but also with online audiences that read scientific posts with varying degrees of understanding, interest, and trust. It also means finding a place in that world for yourself, so you can communicate the things that matter to you, and become the kind of communicator you want to be.

JANET D. STEMWEDEL is a nonpracticing chemist and professor of philosophy at San José State University. Her work focuses on scientific knowledge-building and ethics. She also writes the blog *Adventures in Ethics and Science* and has written for the *Scientific American Blog Network.*

Janet is based in San Jose. Find her at her website, http://www.stemwedel.org, or follow her on Twitter, @docfreeride.

Notes

1. Kenneth D. Pimple, "Six Domains of Research Ethics," *Science and Engineering Ethics* 8, no. 2 (2002): 191–205.

2. Dominique Brossard and Dietram A. Scheufele, "Science, New Media, and the Public," *Science*, January 4, 2013, http://www.sciencemag.org/content/339/6115/40.

3. Mathew Ingram, "Appeals Court Says Blogs Are Not Only Media, They're an Important Source of News and Commentary," *Gigoam*, April 17, 2014, https://gigaom.com/2014/04/17/appeals-court-says-blogs-are-not-only-media-theyre-an-important-source-of-news-and-public-commentary.

4. "SPJ Code of Ethics," Society of Professional Journalists, last modified September 6, 2014, http://www.spj.org/ethicscode.asp.

7

The Deal with Networks

DANIELLE N. LEE

Over time, many science blogs that started as an individual effort have moved to large networks such as those at Scientific American *and* Discover. *But what are the pros and cons of blogging in a network? How does it compare to blogging alone? Danielle Lee, postdoctoral researcher at Cornell University and blogger at Scientific American Blog Network, takes us through the pros and cons of working within a network.*

I began science blogging in 2007. *Urban Science Adventures!* was a labor of love and a personal journal for my developing pedagogical philosophies. At the time, there were no mainstream science television programs that featured women or persons of color as hosts or regular contributors. Digital media was emerging and I wanted to position my blog as an online destination for audiences traditionally underserved by science and science communication.

My long-term blogging goal has always been to access a larger, higher platform so that I can affect conversations pertaining to access and diversity in the sciences. Yet when I began blogging, I was in graduate school completing my dissertation. Trying to find the time and mental energy to create engaging blog posts while juggling my real-life responsibilities of science research and outreach proved to be overwhelming sometimes. As an independent blogger I was responsible for everything, from creating the content to driving traffic to the blog, as well as handling web maintenance. To reach my goal I had to decide which was more important—marketing the blog or blogging.

As blogs became more popular as an alternative news and information source, my interest in blogging for a network increased. Science blogging was such a nascent and dynamic platform; belonging to a network represented a step up in my mind. It signaled that a group of peers recognized the value of a particular voice as a scientist and communicator. A network would be a signal that I was not alone and that my blog was engaging and likely interesting to broader audiences. I would feel honored to be invited to a network, to join a space—even an intellectual one—and to work alongside others whose work and words I admired.

Blogging on a network is akin to becoming a member of a cooperative or a franchise for a popular chain restaurant. In the cooperative example, groups of individuals who share a common mission come together and pool resources (money and audience) to host a blog network. Examples include *Scientopia* and the *Southern Fried Science Network*. They include under the network's banner a roster of well-known blogs, each by strong contributors.

In the franchise example, an established media or publishing company creates a blog network as a new communication offering, both to appeal to their built-in audiences and to attract new and different demographics to the brand. Organizations such as *National Geographic* and *Scientific American* use financial resources and connections to attract strong—and often already popular—bloggers to their organizations. Typically, an opportunity to join one of these networks is offered only after a blogger has already independently demonstrated his or her communication prowess.

I wrote as an independent science blogger for three years before joining the co-operative *Southern Fried Science Blog Network.* While there, I created *SouthernPlayalisticEvolutionMusic,* a blog that explained evolutionary biology with hip-hop songs.[1] One year later, in 2011, I was invited to join the newly established *Scientific American Blog Network.* I accepted because I felt I needed a space that would allow me to develop a mature voice and discuss more advanced biological concepts with more sophisticated readers. I also longed to interact more with other science bloggers. I did not need to join a network to create this new science blog, but allying myself with an emerging cluster of other science blogs felt like the best move for me at the time. I did not want my blog to get lost in the rapidly expanding science-blogging world.

Is it worth joining a blog network? That depends on the goals and strengths of your blog. It is worth a pause if absolute freedom is important to you. Although most networks large and small assure bloggers that they will have autonomy, blog posts can be removed or altered by the network editor. It is worth considering whether you are interested in the increased exposure for your work that comes with joining a network, and whether you would rather focus more on your content (and other real-life science activities) than on managing the blog's logistics and on acquiring traffic. For the independent blogger who has found his or her voice, networked blogging offers several benefits: increased exposure, a certain level of "klout" depending on the network's reputation, access to financial resources (for writers and their blogs), and professional development.

Increased Exposure

When it comes to attracting readers, networks are easier to find in the vast World Wide Web. At the height of popularity of *Urban Science Adventures!*, between spring 2008 and spring 2011, the site averaged eight to ten thousand page views per month. *The Urban Scientist* at *Scientific American* has had more page views in a single year (291,500 from April 1, 2013, to April 30, 2014) than *Urban Science Adventures!* has had in its entire history (284,155 from May 2007 to April 2014).

Networks attract more readers not only from built-in audiences of a brand name publication or organization, but also due to the collective reach of its team of bloggers. With several or even dozens of individuals helping to push out and share each other's content, blog posts are more likely to reach broader audiences. At smaller co-op networks each blogger often automatically tweets out links to all new posts on the network. At larger media-owned networks individual bloggers do the same, while the company also often has a live feed of updates that goes out around the clock. When you consider the cumulative effect of all of these multiple audiences over various social media platforms, then it's easy to see how the reach of a given blog post is generally far greater in a network than when an individual blogger is in charge of distribution.

Blogging within a community can mean support, but it also comes with the risk of getting lost in the crowd. When visitors arrive at the main page of a network they are offered many choices. At smaller and medium-size networks, new visitors are likely to visit several of the blogs. At larger networks, the options may be overwhelming and a reader will only pick a few to visit or only visit his or her favorite blog and move on. In addition, at larger networks bloggers can lose some individuality: in some cases bloggers are no longer referred to by name, but rather become "blogger at (fill in the blank) network." Even more concerning to me is the risk of a networked blog losing its unique voice and appeal. The distinctiveness of some blogs may become washed out as new readers push for a more familiar "scientific tone" from member blogs because of bloggers' affiliation with a traditional science publishing organization.

A Higher Platform for Your Message

There is strength in numbers, and established blog networks can really buoy the careers of nascent bloggers. For networks with recognized media credentials, affiliation can bring additional respectability and credibility to your outreach. Morever, it can make aquiring media credentials to attend and cover scientific conferences and other events easier. Freelancers often have to present several samples of materials that have been published at recognizable outlets in order to get press passes to attend meetings. Affiliation with a publishing company network helps bloggers overcome this hurdle.

Deeper Pockets

Blogging is fun and thanks to easy-to-use platforms, starting a blog is relatively straightforward. But glitches still occur. When I was in Tanzania conducting field research, I was not able to post reliably; my Internet access was spotty. Because I was part of a network, my blog editor was able to upload pictures on my behalf and format my posts as needed. I was able to continue posting dynamic pieces with the help of the team I had available to assist me.

Networks are also more advantageous than independently owned blogs when it comes to fixing major maintence issues. In the summer of 2012, the *Fab Lab* with Crazy Aunt Lindsey was on its way to becoming the hottest science outreach video series program on the web.[2] "Crazy Aunt" Lindsey Murphy had recently completed shooting season three of her web show, which was hosted at her own domain. Along with amazing blog posts about after-school science activities, all of her other content was compromised when her site was hacked. Although she had backed up the content through her second season, her newest and most creative content was lost: repairing the server was just too expensive. "When it's just you responsible for everything—the content, the writing, the performing, the hosting—you realize it's so many different responsibilities," Murphy explained.

Smaller co-op network communities often appoint someone or pay a third party to address these types of issues to prevent a loss of content. Larger networks managed by companies usually have dedicated personnel to pre-empt and address major issues like these. Having someone—or even better, a team—with expertise in solving such problems is a time- and mind-saver.

Protection from random viruses or deliberate attacks is especially important at blog networks with high levels of traffic. For networks like *Freethought Blogs* that routinely blog about incendiary topics, having extra heavy security and web support is essential. During Ian Cromwell's time at Freethought Blogs (writing his *"Crommunist" Manifesto*), he recalls five or six significant attacks on the network that affected availability.

Financial Opportunities

As flattering as an invite to blog for a network may be, it does not usually mean a big payday. While there is a chance you may be offered money, payment is certainly not universal. Science blogging networks backed by large publishing and media companies (for example, *Scientific American, Popular Science, Discover,* and *National Geographic*) usually pay regular contributors. Some of the smaller, emergent networks (such as *Double X Science*) pay contributors when they can. A more likely outcome is an increased opportunity to contribute freelance pieces to other outlets and to speak to wider audiences, by virtue of having a more prominent platform. A freelancer can earn about the same amount of money for a single article submitted to general audience magazines and news websites (like *Scientific American* or Ebony.com) as a network blogger earns for four or more blog posts per month.

Although the pay earned for blog posts may seem exploitative to some, the benefit of having a blog hosted by a major network is that it is easier to pitch to other outlets. Name-dropping an association with a popular network comes in handy when pitching to editors elsewhere. It becomes easier to find work and connect with editors at news websites that often commission longer, more in-depth pieces and that pay more, too (like *Slate* or *Mother Jones*) by being connected with a media brand that lends credibility to your work.

Blogging at a major network increased my visibility to the extent that I now receive invitations several times a year to speak about blogging and science communication at conferences and at universities as well as to give talks about my research. Many of these engagements come with money to cover my travel and lodging costs plus an honorarium, but not all them. Be mindful that not all invitations to share your expertise—via writing or in person—will offer compensation. Eager, novice writers must be careful to balance the prospect of increased exposure against the risk of professional exploitation.

Professional Development

I began as a niche blogger: as an African American science blogger I was one of just a few science bloggers from an underrepresented minority group. Since joining a network I have become recognized as a professional writer and an advocate for diversity in science communication. Since I write for a network with cachet among those active in publishing and other media, I can leverage my proximity to a popular brand, in particular a brand that is all about science, to influence science media delivery to underserved audiences. I have participated in discussions concerning diversity in science communication with members of the American Association for the Advancement of Science, the National Association of Science Writers, and the National Association of Black Journalists.

Bringing individual writers together under one tent creates multiple networking opportunities for everyone. Especially if there are heavy hitters in the network, the opportunities for young writers and science outreach professionals are immense. In addition, do not underestimate the benefit of having access to editors. They can open doors. For example, in 2013 I was able to organize the first science journalism workshop for the National Association of Black Journalists. Thanks to our shared affiliation, *Scientific American* news editor Robin Lloyd volunteered to attend the meeting as one of the featured panelists.

Finally, I found that belonging to a network, where I was digitally flanked by other amazing scientists and science writers, motivated me to step up my own game. I am more thoughtful about my posts. I research my hard-hitting topics more carefully. I am more conscientious of image appropriation and links I use in my posts. And I am more mindful of copyright and legal infringement risks—of my own material and that of others—than when I was blogging independently.

My blogging has matured, as have my ideas about professional science outreach and communication, since joining a network. I doubt that my writing or my online activism would have grown in such a way if I had continued to blog independently. Specifically, my ideas concerning feminism, diversity, and inclusion in science and science education have been positively influenced by my blog network peers. I compare the topics I cover to those of others in the network, and I consider the tags used for each post. Although I may speak on certain issues or offer a particular perspective, I want to be sure that my content is complementary to other conversations already occuring on the network and within the field of science communication in general.

When your blog stands alongside other blogs that provide thought-provoking and engaging content, it inspires you to aspire. When you read or view dynamic pieces on your network's front page, you feel motivated to bring your best to the table: to your colleagues and to all of your readers. Blogging on a network has not been without its pitfalls, but affiliation with a high-profile, commonly recognized media brand has been a boon for me professionally, and for my message.

DANIELLE N. LEE is a postdoctoral researcher at Cornell University. She also runs the blog *The Urban Scientist* at the *Scientific American Blog Network*. Danielle was named one of the White House Champions of Change in STEM Diversity and Access in 2014, is one of the 2014 Ebony Power 100, and is a 2015 TED fellow.

Danielle is based in Ithaca, N.Y. Find her at her website, http://about.me/DNLee, or follow her on Twitter, @DNLee5.

Notes

1. Danielle N. Lee, "And a New Science Blog Is Born . . . SouthernPlayalistic EvolutionMusic," *The Urban Scientist,* September 15, 2010, http://urban-science.blogspot.com/2010/09/and-new-science-blog-is.html.

2. Lindsey Murphy, "UPDATED: Crazy Aunt Lindsey and the Big Bang Theory (Also: How Not to Lose It All Online)," Lindsey Inc., http://lindseymurphy.com/crazy-aunt-lindsey-and-the-big-bang-theory-also-how-to-not-lose-it-all-online.

8

Indie Blogging: On Being a "Ronin"

ZEN FAULKES

Over time, many science blogs have moved from individual and personal blogs to large networks such as those at Scientific American *and* Discover. *But what are the pros and cons of blogging in a network? How does it compare to blogging alone? Zen Faulkes, a tenured professor of biology at University of Texas Pan American and an "indie" blogger, takes us through the pros and cons of blogging alone.*

The renown of samurai, the warriors of Japan's feudal era, has long outlasted the age in which they lived, and spread beyond the shores of their island. A samurai was a member of Japanese nobility who was expected to be both a warrior and a scholar. The samurai code of honor demanded a willingness to die for the warriors' lord at any instant. Some of the most famous tales of samurai culture revolve around *ronin*. Ronin were samurai who had no allegiance to a lord, either because the samurai had fallen out of favor with his master, or because his master had died. The tale of forty-seven ronin who avenge the murder of their master is so famous that it made the jump from Japanese literature to a big-budget, special effects–laden Hollywood movie.

To Western ears, ronin have a certain mystique, even by samurai standards. Having no master implies freedom and independence. You can do what you want. You aren't working for "the man." But a closer look shows that a ronin's lot is far from rosy. Writer John Wick has explored samurai culture in several of his projects, such as *Legend of the Five Rings*. He shows that ronin are, essentially, homeless people. Citing *Yojimbo* and *The Seven Samurai* as examples, he wrote, "They have shitty equipment, shitty clothes, they have to beg for food and get ridiculed by 'real' samurai for being cowards. The whole point of being a ronin is begging for a living."

Both of these views evoke an experience similar to that of being an independent blogger, one who is unaffiliated with any of the "official" blogging networks. You have freedom, but that means you are on your own. Being entirely on your own can feel like begging, albeit for attention rather than food. But as stories of the forty-seven ronin prove, independent blogging does not mean ending up in obscurity.

New science bloggers starting today might well be leery of being entirely on their own. They might view independent blogging as a situation to be tolerated for as short a time as possible, a stepping-stone to an invitation to blog for a network. Networks have been a huge part of the science blogosphere since the heyday of *ScienceBlogs*, roughly 2006 to 2010. Many bloggers left *ScienceBlogs* in the summer of 2010, but the concept of a blogging network had been so successful that many other science blogging networks were set up. Several bloggers have switched networks multiple times, and it's clear that networks are actively recruiting certain desirable bloggers to their sites. Several science blogging networks are linked to famous magazine and media brands like *National Geographic, Scientific American, Wired,* and *Discover.* Print may be dying, but the recognition of those well-known names still carries a lot of weight, and being associated with those names means a bigger potential audience.

In other words, it would be easy to see independent blogging as just an audition. Some see the measure of real success in science blogging as "getting picked" for a network. This is a familiar situation for a lot of prospective science bloggers. Bloggers who come from an academic background have spent their careers waiting to be picked: by universities, by doctoral and postdoc supervisors, by search committees, by the top journals, and by funding agencies. Bloggers from a writing background also have careers that can depend on getting picked by publishers, editors, and media outlets. People will often take low- or no-paying work at bigger outlets for the chance to raise their public profile.

This situation is, weirdly, an inversion of blogging's beginning. Blogging started as push-button publishing for the people. It was a way to share writing with anyone, anywhere, without having to be "picked" first. The move to networks is probably a logical progression of the maturation and professionalization of the blogosphere, but when I began blogging, I couldn't aspire to join a network, because there were no networks to join.

I currently have three active science-related blogs. My first, which I started in 2002, I later named *NeuroDojo.blogspot.com;* next were *Marmorkrebs.blogspot.com* in 2007 and *Better Posters.blogspot.com* in 2008. Since my start in 2002, I have never been part of a network. I did apply to join the *ScienceBlogs* network at one point, but I was rightly rejected. *NeuroDojo* was not ready, and only later got "born again hard" as a science blog. Sometimes I did feel I was on the periphery rather than "in the game," but now I have fully embraced my status as a ronin blogger.

There are advantages to belonging to a network, but they may not be as great as you think. There is a certain visibility that comes with a network, but being on a network in and of itself does not guarantee readership. Ultimately, networks function best when they create and facilitate communities. To the extent that I have been successful as a blogger—whether measured in page views or that people know who I am and what I do—part of that success is because I embraced the social nature of the blogosphere. I worked to join existing communities.

A key moment in the development of *NeuroDojo* was when I put posts on a science blogging aggregator, *bpr3.org* (Blogging on Peer Reviewed Research Reporting; this became *ResearchBlogging.org*, which has been superseded by *ScienceSeeker.org*). All three of these websites compiled scientific blog posts, particularly about journal articles, into a single "one stop shop," sorted by discipline. You could find math blog posts, astronomy blog posts, or biology blog posts. These aggregating sites provided new readers with a useful key to the diffuse science blogosphere. I noticed a bump in visitors to my blog when my posts went on bpr3.org. Prior to this, *NeuroDojo* had been a very inward-looking blog, very much about my own experiences as a working academic. Posting on bpr3.org shifted my focus outward. I started writing more posts about research papers, because those could be syndicated through the aggregator. The hits slowly started going up. This was my first inkling of an important realization: ronin may have no master, but they still belong within a community.

I created the *Marmorkrebs* blog with a community in mind: a small, scientific research community that was interested in a single organism—the all-female line of crayfish, *Marmorkrebs*—which gives the blog its name. This was a very niche project, but I know it has reached the intended audience. Other researchers in the field send me materials for the blog, and the blog has even been mentioned in a few scientific papers.

By far, the *Better Posters* blog is my most successful online project. It's also the project that taught me about the importance of the social nature of a blog. The mission of the *Better Posters* blog is right there in the name: "better." It was created to be a helpful resource for others, because constant improvement is the samurai way.

I didn't realize it when I started, but that goal of providing aid made the blog stand out on the Internet. You see, certain things are overrepresented on the Internet. Near the top of that list—just slightly below funny cats, naked women, and maybe bacon—is snark. We geeks love to snark. It fits into a deep-seated need to show how smart we are. We want everyone to know not just when something is *wrong* on the Internet, but that we saw it first. Both good scientists and good journalists are trained to be critical, and learning to deliver critical appraisals of evidence often manifests itself as snark. Viral pictures making dubious claims that are retweeted and liked on social media practically beg for snarky ripostes and takedowns. In an Internet filled with snark, something like the *Better Posters* blog, which has the avowed goal of continual improvement, stands out.

The way to achieve continual improvement and the elimination of errors often does mean criticism. But criticism does not have to involve snark. I have often critiqued on the blog poster presentations that I found on conference websites. These were often award-winning posters, which I suspect won because their science was excellent, though their design often left a lot to be desired. Readers seemed to realize that my critiques were intended not to mock or laugh at these posters (samurai have no reason to be cruel), but to help others avoid repeating mistakes. Because of that, people started to send me their conference posters, unbidden. I wrote that I would accept posters for critique if I could feature them on the blog. Readers thus became contributors.

The importance of community was again brought home to me when people started recommending *Better Posters,* mostly on Twitter. The recommendations were generally not for individual posts, but for the whole body of work on the blog, and I started compiling these on a "What people are saying" sidebar. These served as the blog's social proof. The Internet shows us that we like what other people like. No amount of self-promotion and retweeting your own blog will ever have the power of someone else recommending your blog. The biggest jumps in traffic were driven by recommendations by other people, not my own self-promotion. That's the power of social proof.

Your early readers are important, because they can act as your amplifiers. Today one of the best ways to find those first readers is on Twitter—but Google+, Facebook, and other social networks could be catching up fast. Regardless of where your readers come from, remember to thank, whenever possible, the people who recommend your work.

There are many other ways to interact with the larger community as a ronin blogger. Write posts that are reactions to other people's posts. Do not be afraid to write about the same stuff other people are writing about, like the latest scientific controversy. Just make sure that you have your own voice and your own angle. Comment on other people's blogs. Link out to related posts—a lot. And interact with people not just on blogs, but also on lots of other forms of social media, like Twitter, Facebook, Pinterest, or whatever strikes your fancy. A ronin need not be a hermit.

Communities are built on common interests. The more shared interests, or the deeper those shared interests, the stronger and tighter the community. Consequently, you may be able to build a community more readily if you identify what those interests are. For example, *Better Posters* has a very tight focus: everything in the blog is related to conference posters in some way. I sometimes think the smartest thing I ever did in blogging was to split off my writing about conference posters and put it into its own blog. I could have written posts on my general blog and tagged the posts with "conference posters." I do this with posts about oral presentations, which I do include on my general blog. Those posts are popular, and I get good feedback on them, but it is nowhere near the attention that the posters blog gets. You are much more likely to get attention if you can find a single, unoccupied niche and exploit it.

The focused subject helps build readership, for two reasons. First, a specialty "boutique shop" creates clarity of purpose. Audiences want to know what to expect. As we learned from the *Blues Brothers* movie, people who go to a country bar will start throwing their beer bottles if the band does not play country music. Focus means that readers know what they are getting if they read the blog, and they are not put off by topics they don't like. With "everything plus the kitchen sink" blogs, there is a risk of putting people off if they come for the science but get a lot of posts about other topics in between.

Second, the tight focus makes it easy for people to recommend the blog. "If you want to learn about conference posters, this is the blog you need to visit" is a much more straightforward instruction for one reader to give another than "Go to this blog and search for the posts tagged 'posters.' Or was it 'conference posters'? Or maybe 'poster design'?"

Having spent most of this chapter emphasizing community, social proof, and networking, why do I remain a ronin blogger?

First, I have enjoyed consistency. My blogs have never moved to new websites with new URLs. In the long term, this helps people to find my blogs. Thanks to people typing questions into search engines, some posts continue to attract traffic long after I posted them, and it is easier for people to find old posts by lucky searches because the link to those individual posts has never changed. My own home page provides a "portal" to my blogs, and allows people who find one of my projects to track back to find others.

Second, nobody can take down posts but me. Although blog networks often seem to give their writers considerable freedom, it's not always clear how far the leash will stretch. Blogger Danielle Lee, for instance, once had a post from her blog, *The Urban Scientist*, temporarily removed by editorial staff on the Scientific American network.[1] The case attracted a lot of attention because it was extremely unusual. Still, it showed that blog networks, particularly those sponsored by legacy print media, might be heavy-handed in their efforts to protect their brands. Some bloggers have left networks and gone back to blogging on their own, in part because the desired support from the network leaders was not there. Serving a master means there is room for conflict, and may lead to a lingering fear that your superior might not always agree with you.

Finally, while there's no denying that it's nice to have an audience, I blog for myself first. I have a career that does not depend on blogging, so I have no need to chase page views. I do not need to run ads that offer "rules for a flat stomach!" to subsidize my writing. I don't need anyone to choose my writing or to pass an audition.

Despite the challenges of being a ronin blogger, then, there are many things I savor. I enjoy that my blogs have continuity, and that a reader can crawl through a blog in its entirety at one single web address. I've added new themes and threads to my blogs without worrying that I was stepping outside the boundaries of what that blog is "supposed" to be about. My blogging course is not buffeted this way or that by winds of chance. Because of why I blog, I choose to stay a ronin blogger.

ZEN FAULKES is a neuroethologist at the University of Texas Rio Grande Valley. He runs the blogs *Neurodojo* and the *Better Posters* blog. He is also author of the eBook *Presentation Tips.*

Zen is based in Edinburg, Texas. Find him at his website, http://doctorzen.net, or follow him on Twitter, @doctorzen.

Note

1. Danielle Lee, "Responding to No Name Life Science Blog Editor Who Called Me Out of My Name," *The Urban Scientist* (blog), *Scientific American,* October 11, 2013, http://blogs.scientificamerican.com/urban-scientist/2013/10/11/give-trouble-to-others-but-not-me.

9
Getting Interactive

ROSE EVELETH

Many people take up science blogging because they love words. But on the Internet, words aren't everything. Rose Eveleth, freelance writer and producer, explains how interactive elements can bring your simple blogging up a notch.

You've had to write things down for most of your life. You've been evaluated based on your writing for probably as long as you can remember—from spelling tests to cover letters to lab reports to grants. And you're probably interested in blogging because you actually like writing. You like words. Words are your escape, your weapons against the straight lines of test tubes and board meetings in the fight for sanity. So when it comes time to blog, your words probably come out first in the form of, well, words. Text on a glowing screen. Pretty, isn't it? Blogging for most people is simply a new kind of writing. But it can be more. Creating and using nonword elements in your blog will help you make your point, engage readers, and have even more fun.

Now let me do the very odd thing of describing to you, in words, why, how, and when you might want to not use words. Here goes nothing.

Why Should I Care?

Let's start with why. Why not just say what you mean? Why not just write it all out? Don't get me wrong, words are awesome. Rudyard Kipling once wrote (in words), "Words are, of course, the most powerful drug used by mankind." But let's imagine something else for a second.

Let's say you're really into speed skating. Speed skating is this incredible sport where people put on spandex and strap giant metal blades to their feet and whip themselves around a track covered in ice. And I mean whip—these people are going incredibly quickly. In the 2010 Winter Olympics, Christine Nesbitt of Canada won gold by skating a thousand meters in 1:16.56. That's one minute, sixteen seconds, and some change. That change matters here, though, because right on her tail was Annette Gerritsen, the silver medalist who crossed the finish line at 1:16.58. The difference between gold and silver was .02 seconds. How fast is that? In words, that's faster than it takes for a hummingbird to flap its wings. But do you really know how fast that is? Can you imagine, in your head, a hummingbird? Are you counting its imaginary wing beats and looking at your imaginary watch?

What if we tried to explain that story in another way, a nonwordy way. That's what Amanda Cox at the *New York Times* did. She represented each of those finishes with a little ping.[1] When you play the finish back and hear those two little pings so, so close to one another, you really, truly understand what it means to finish .02 seconds behind the gold medalist.

In 2010, *Radiolab* producer Jad Abumrad got up on stage at PopTech and played a sound.[2] It was the sound of his printer freaking out. It sounds kind of like "click srrrrrr schhhhhhh beep eeeee srrrr schhhh ft ft ft ft ee ee ee ee." Except it doesn't really sound like that at all, because I've just given you a bunch of letters smushed together in a feeble attempt to re-create a sound. Abumrad was talking about the sound of failure, of things breaking, of the world around you suddenly jolting to a halt, breaking a nice, pretty flow that we've designed for ourselves. He could have said, "One time, my printer made a really weird noise." He could have even shown you a bunch of smooshed up letters like I did. But instead he gave us that real sound, the sound of gears and belts and whatever else is in a printer that usually co-operates harmoniously.

Or maybe you're covering climate change and you want to discuss how the environment has changed over time. You could say "the Gangotri glacier has receded 850 meters" or you could stitch together several images from NASA into a GIF to show exactly what that looks like. I have another speed skating example, but I think by now you get what I'm writing about so I'll spare you.

All of these examples show that sometimes the best way to tell a story involves more than just words. But if you're the kind of person who measures self-worth in social media data, think of it this way. By some measures, tweets with images in them get twice the amount of engagement as those without. Posts with pictures on Facebook get 53 percent more likes than posts without. Videos go viral far more frequently than long strings of text. *The Crazy Nastyass Honey Badger* video—a film I consider to be a totally legitimate form of science communication—has nearly 70 million views. People like pretty pictures, things they can click on, and stuff that bounces around in front of them. Just don't make it autoplay, for the love of all things science.

Which Shiny Internet Things Should I Try?

If you keep thinking to yourself "You keep saying that I should include sounds and video and infographics, but how do I make this stuff?" then this is your section. This is the part where I'll list some—but certainly not all—of the various media and tools you can use to play with your content.

But before I do that, I'm going to give you the One True Rule of making things: make them. I know it sounds dumb, but you're never going to make anything until you actually make something. The first things you make will probably be bad. That's okay. How many times did the Wright brothers crash their weird wing contraptions before they made a plane? A lot. Had they been on the Internet there might have even been a lot of mean-spirited people making fun of them about it. But I'm editing this chapter on a plane that has an upstairs, so you and I both know who had the last laugh there.

Okay, now that you know the One True Rule of making things, here are some (but not all) of the things you could consider making.

IMAGES

These are easy, you know about them, and there are other chapters in this book that will tell you all about them. Just remember to credit your photographers and artists. It matters.

GIFS

Beyond just the silly "cat dressed as a shark riding a Roomba chasing a duckling" GIFs (yes, that exists), these little movies can be really useful to explain things or illustrate points.[3] You can make a GIF by stitching literally any images together, including ones you've drawn yourself. Are you discussing a live map that updates every few minutes? Make a GIF of a few seconds of that map moving about. Are you describing how awesome solar flares are? Include a GIF of a solar flare erupting! Explaining how sperm swim? Sperm GIF! Speed skating? GIF away, my friends! You get the idea. There are a lot of ways to make GIFs, including Photoshop and a variety of websites like http://makeagif.com.

SOUNDS

You can find all sorts of sound on the Internet, at free sound archives like Free Sound, the Free Music Archive, and the British Library's sound collection (http://sounds.bl.uk). Or you can record your own and upload them to one of the multitude of sound-hosting websites.

ANIMATIONS OR VIDEOS

Animations might seem daunting to make, and they definitely take time, but they can also be quick and easy. Take Vine, for example. Move a few things around in front of your cell phone's camera and you've got an animation. Want to show how cells divide, how molecules move, or how far away the moon is from the Earth? Try delivering the scale and motion visually with an animation or video.[4]

TIMELINES

There are all sorts of online tools that you can use to create a nice, visual, interactive timeline to depict the course of events.[5] Maybe you're covering something with a long history or something that happened very quickly, and you want to break down the different events. Using a timeline rather than describing the series of events in words can help readers visualize the sequence—whether that's the sequence of the Ebola outbreak or the chain of events that led to the Arab Spring uprising.[6]

These are just a few things on the endless list of things you can do that aren't text based. You could include quizzes, discussions, tweet chats, infographics, podcasts, and more. It all just depends on what you're trying to do.

When Should I Use a Shiny Internet Thing?

A recent *Onion* headline read "Internet Users Demand Less Interactivity." "Speaking with reporters, web users expressed a near unanimous desire to visit a website and simply look at it, for once, without having every aspect of the user interface tailored to a set of demographic information culled from their previous browsing history. In addition, citizens overwhelmingly voiced their wish for a straightforward one-way conduit of information, and specifically one that did not require any kind of participation on their part," the authors wrote. And it's true, sometimes you don't want a million GIFs to just hear about some new study that really has nothing to do with that GIF of a cat trying to jump off the couch and failing.

As Uncle Ben tells Spiderman, "With great power comes great responsibility." Just because you can make your blog the most sparkly sound- and video-filled thing around doesn't mean you should.

"But Rose," you're thinking, "I thought the One True Rule was to make things!"

Yes, it is. But making and publishing aren't always the same. If you make something, and realize that it's not helping, it's confusing, or it's simply embarrassing, you haven't wasted your time. You've still learned how to make something. You've learned how to use a particular piece of software or an app. That something might not be right for the project you're working on at that moment, but chances are it—or something like it—will be useful to you in the future. At the very least you've forced your brain to come up with something it wouldn't have normally come up with. Such as a unique perspective on speed skating.

Just like most things in the world, it comes down to discretion. And you already probably have pretty good discretion to begin with. You're a writer, remember? You're good at telling stories with words. This really isn't so different. Think of your visual elements as being like examples in your story. Does the example support the point you're trying to make? Does it add something new that other examples failed to provide? Is it interesting? Are people going to remember it? Does it require a ton of extra explanation to set up?

These are all questions you'd ask of any written example in a blog post or story. Those that make sense, fit nicely into your story, and add value for your readers get to stay and hang out. Those that require lots of explanation, that are confusing, or that don't add much are banished to the trash.

But wait! Before you trash it, think again about how to fix it. Why isn't it working? What's the confusing part? Okay, sometimes you have to just trash it. But that's all right. Remember, you lose 100 percent of the speed skating races you don't enter. Trying and failing sometimes is never as bad as not trying at all.

Listen, making stuff for the Internet—words or otherwise—is hard. A lot of what you make is probably going to be mediocre. Some of it might even be bad. But some of it will be good and all of it will be useful—at least to yourself as a way to learn how. Try using sound to illustrate your point. Try putting some images together into a comic form or a timeline or a quick animation. Try a podcast in which you interview yourself. Try everything. I'll write that again, because it's important. Perhaps the most important. Try everything. Remember the One True Rule? Good, now go make stuff.

ROSE EVELETH is a freelance producer, designer, writer, and animator. She has written and produced for the *Atlantic, BBC Future, Nautilus Magazine, NOVA,* and others. She is also the founder of *Science Studio,* a place for science multimedia, and an editor at *Story Collider,* a science podcast.

Rose is based in Brooklyn, N.Y. Find her on her website, http://www.roseveleth.com, or follow her on Twitter, @roseveleth.

Notes

1. Amanda Cox, "Fractions of a Second: An Olympic Musical," *New York Times,* February 26, 2010, http://www.nytimes.com/interactive/2010/02/26/sports/olympics/20100226-olysymphony.html.

2. Jad Abumrad, "Sound and Science with Jad Abumrad," filmed 2010, PopTech video, http://poptech.org/popcasts/sound_and_science_with_jad_abumrad.

3. Orbitn, post on Imgur, http://imgur.com/gallery/IgobS4F.

4. "The Animated Life of A. R. Wallace (Director's Cut)," Vimeo video, 7:55, posted by "Sweet Fern Productions," 2014, http://www.chicagomanualofstyle.org/16/ch14/ch14_sec280.html.

5. "Timeline," Northwestern University Knight Lab, 2013, http://timeline.knightlab.com.

6. Garry Blight, Sheila Pulham, and Paul Torpey, "Arab Spring: An Interactive Timeline of Middle East Protests," *The Guardian,* January 5, 2012, http://www.theguardian.com/world/interactive/2011/mar/22/middle-east-protest-interactive-timeline.

10

Brevity Is the Soul of Microblogging

JOE HANSON

When people think of blogging, they may think of thousands of words on a single page. But there's a lot to be said for keeping it short. Joe Hanson, the mind behind the popular YouTube and Tumblr series It's Okay To Be Smart, *describes the important role that microblogs can play in your science communication efforts.*

You might be a Whovian, a Nerdfighter, or a Potterhead. Perhaps you enjoy *Garfield* comics without Garfield, or when *Family Circus* is captioned with quotes by Friedrich Nietzsche. Heck, maybe you just want to look at the same picture of *Full House* actor Dave Coulier every day.

These are all fandoms, and they all live on microblogs like Tumblr.

A fandom is a community of cultural camaraderie driven by shared interests. Today microblogs are the meeting places where these groups share those interests, whether mainstream or ridiculously obscure. Like many communities, fandoms use a shared language to identify insiders, a shorthand syntax where one "ships" people, not parcels, and where passionate debates occur over the correct pronunciation of "GIF." And above all else, the primary currency is the "share."

Microblogs are invaluable to pop culture, and perfect for a teenager's thought diary, but they are still rarely used for science communication. On its own, a microblog isn't best suited for carrying out a broad, institutional science communication effort. But they can be important parts of those larger efforts, useful tools for doing a very particular kind of science communication directed at a very particular audience. Current users of microblog platforms are young, spend a lot of time online, are well-educated, and are fairly evenly split between males and females.[1] Tumblr holds a reputation for engaging groups typically underrepresented in science, especially women, racial minorities, and people identifying as LGBTQ. Tumblr users are the proverbial "key demographic," that is, replete with important science fandoms waiting to be noticed.

Microblogs trace their popularity to ease of use. If we do a basic etymological dissection, "microblogging" is really just blogging, only divided into smaller units of content, delivered and consumed in lesser units of time. When compared with its "macroblogging" relatives, microblogging distinguishes itself in another important way. It has become a "broadcasting" medium.

Today, almost without exception, microblogging platforms integrate the core elements of social networks into their architecture, most notably the "follow," "like," and "share" actions. Indeed, it's becoming difficult to see where something stops being a microblogging platform and becomes a social network, and vice versa.

Microblogging could be defined, if we're forced to do so, as the blog, distilled to its individually shareable units (items such as photos, videos, links, and bits of text) and broadcast throughout a social network. Twitter fits that definition, and is often referred to as a microblogging platform. But I've chosen to omit it from this discussion, because it has evolved into a broadcast medium in a class of its own. Microblogging, then, is social blogging.

As it exists today, microblogging is dominated by one platform: Tumblr. Other platforms exist, notably Pinterest, but Tumblr's diverse content and richer history make it an ample test case. We can understand a great deal about Tumblr's utility, and therefore the more general usefulness of microblogging, by studying its genesis.

Microblogging-like activity traces back in one form or another to the earliest days of the World Wide Web, and probably even before that. In 2005, though, then-seventeen-year-old Christian Neukirchen's anarchaia.org was the first site described, rather presciently, as a "tumblelog." It was so named because a reader could quickly "tumble" down the page, a column stacked in reverse chronological order filled with links, images, and other short, free-form content.[2] Later that year, blogger Jason Kottke described the new tumblelogs as "a quick and dirty stream of consciousness" and "really just a way to quickly publish the 'stuff' that you run across every day on the web."[3]

David Karp, with the help of Marco Arment (later of Instapaper fame), built and launched the first tumblelogging platform in 2007. Tumblr was based on Marcel Molina's hand-coded tumblelog *Projectionist* (http://project.ioni.st), which set the style, organization, and format for nearly every tumblelog that came after it. In the modern tumblelog format, different types of content are displayed in different ways, and the focus of a post is shifted from standard blocks of original text to individual content units: links, snippets of text, images, videos, or even audio. Today, much like Kleenex or

Band-Aids, these tumblelogs have become so eponymous with the platform that hosts them that we simply call them "tumblrs."

The evolution of Tumblr and other microblogs seems in no small part to be a reaction to traditional blogs' magazine-like and text-heavy format. Often only a small portion of the content on most Tumblrs is created by the person who runs the Tumblr. Back in 2005, Kottke described tumblelogging as feeling closer to editing and less like writing or punditry. David Karp said one of his primary motivations for starting Tumblr was that he "doesn't enjoy writing."[4] Tumblr, and microblogging in general, seem to have made "curation" the name of the game.

Much later, flagship social networks like Facebook and Google+ integrated microblogging into their platforms. What is a status update, or link with comment, if not a microblog post?

Tumblr is good at communicating *any* media or content, be it science, nail art, or *Sherlock* fan fiction. Tumblr is a campground where passionate people reside, and it succeeds in its ability to unite them around very specific campfires to tell very concise campfire stories. Quick, targeted, and passionate.

That being said, there are certain best practices and particular types of media that flourish on Tumblr. But to understand what and why, we have to pin Tumblr to the surgical tray and do a bit of dissection.

Tumblr is a beast with two faces. One is a face that it shows to the outside world: a standard-format web page that resides at a URL and can be customized with all of the standard HTML trappings, albeit within certain limits imposed by the platform. Everyone who signs up for Tumblr has a public blog, whether or not they choose to put anything on it, but that's not really Tumblr any more than a person's Twitter profile page "is" Twitter. Tumblr's true face is the Dashboard.

Only users who are signed up for and signed in to Tumblr can see the Dashboard. Prominently displayed at the top are the seven types of posts: text, photo, quote, link, chat, audio, and video. Clicking one brings up a post editor specifically tailored to that type of content. Beneath that is the post feed, which serves up content in reverse chronological order from all the other Tumblr pages the user "follows." Alongside the post feed is an information panel containing many standard blog tools, including a summary of recent activity on a user's posts, navigation to other Tumblr blogs run by the same user (Tumblr offers multi-author support), the queue of scheduled posts, and a list of followers.

This is the only place that a follower count appears on Tumblr. Unlike on Twitter or Facebook or Google+, Tumblr follower counts are not publicly viewable. This was a deliberate decision on Karp's part. He called such markers of social popularity "really gross."[5] This content-over-fame vibe permeates the network. Perhaps more than on any platform, institutions and brands that want to communicate science must act and speak as part of the community, not down to it.

Photo posts are by far the most popular, with one analysis showing that photos made up 83 percent of posts.[6] Multi-photo posts can be organized in a number of ways. There is also native support for animated GIFs, and Tumblr is primarily responsible for the resurgence of this file format.

Tumblr's design and architecture promote positive interaction among its users, from the choice of a heart icon to denote a "like" to the fact that instead of commenting on a post, users must reblog it to their own pages, which means that all comments will also reside on their pages and become attached to their identities. This strikes me as one of the most important, yet oft-overlooked aspects of the Tumblr microblogging experience, and it maximizes, although does not guarantee, positivity.

The content must be interesting, but interesting content is only part of the recipe. More importantly, microblogging for science requires both effective use of the platform's design and molding the content to the desires and tendencies of the audience. Content, like a Tumblr user, must be quick, targeted, and passionate.

Images are more shareable than text, and they make people pause, if just for a moment. This is not to say that images carry more significance, but the intellectual value of a long text post must be weighed against the very different sort of impact that an image post can have. On Tumblr, even text converted *into* an image has more potential impact than text alone. The once-maligned GIF format has become an art form unto itself, part irony and part awesome. There's something about a moving, silent image, absorbed in mere seconds and with no need for reader intervention, that just plain works.

If you want a post to be seen, liked, and shared, you must tag it. Tumblr's tagging feature is its greatest discovery tool. Posts are accompanied by hashtag-like tags inserted by the user, some broad (like "science") and some oddly specific (like "cats in spacesuits"). On Tumblr, any tag can be tracked, or subscribed to, allowing a user to see any post on Tumblr containing that tag. Popular tags like "science" have tag editors chosen from the Tumblr community that can "feature" a post on the larger, curated tag pages, driving traffic and increasing engagement. Untagged posts are invisible posts.

There are several great science Tumblrs that you can learn from. *Fuck Yeah Fluid Dynamics* (http://fuckyeahfluiddynamics.tumblr.com) serves up GIFs, short videos, and photos illustrating principles of fluid dynamics. Concise explanations are attached below, but are secondary to the truly shareable item: the image or video.

Explore.noodle.org serves up mostly visual fare in subjects ranging from literature to history to science, utilizing Tumblr's powerful tagging system both to organize the posts on the public-facing page, and to target content to Tumblr's countless communities of shared interests.

Brookhaven National Labs runs an institutional microblog (http://brookhavenlab.tumblr.com) that features odd historical archives as well as notable videos and images relating to current Brookhaven research. *Skunkbear* is a microblog (http://skunkbear.tumblr.com), run by NPR's science desk, that publishes original multimedia web journalism specifically produced for Tumblr. On both, posts usually link back to the brand or institute's larger website, where readers can encounter more traditional science communications like press releases and feature journalism.

The ease of sharing on platforms like Tumblr has turned information consumers into curators, with creators often suffering financially. While Tumblr has integrated many elements into its platform to aid in attribution and linking back to the original content creators, the responsibility to engage in these best practices ultimately rests on the shoulders of the end user. Even a cursory tour through your typical Tumblr page indicates that attribution behavior has much room for improvement. But free sharing of media is a cornerstone of the social web, and the protection of intellectual property must find a way to coexist with our sharing culture, hopefully through education and proper behavior on the part of those with influence. The act of curation allows an information consumer to personalize his or her experience, to belong to passionate communities of shared interests with minimal investment of attention. It forces creators to find ways to belong to those communities, rather than to simply deliver content to them. As Tumblr's Danielle Sterle says, "It's a magic river of Internet awesome that you have made all for yourself."[7]

Tumblr blogs have spawned more than a hundred book deals; musicians have released albums directly to their fans with Tumblr;

and photographers, artists, and fashion journalists have used their Tumblr communities to launch many private companies and major brand partnerships.[8] Nerdfighters, the fan community based around Hank and John Green's YouTube projects, has raised hundreds of thousands of dollars for global charities (or, as its members put it, the Foundation to Decrease Worldsuck, http://www.projectforawesome.com). My own Tumblr, *It's Okay To Be Smart,* has allowed me to build a community of hundreds of thousands of followers, to launch a weekly science education video series produced by PBS, and to reach millions of people who wouldn't usually seek out science in their daily lives.

Like every platform or service in the digital toolbox, Tumblr is a very specific hammer for a very specific kind of nail. There is no one-size-fits-all solution to the challenges that science communication faces, but there is also no one kind of audience.

Microblogs may not ever exist as cornerstones of science communication efforts, but they can be an important tool in the creation of science-interested communities. There are few, if any, social media tools today that foster sharing and communicate shared interests as well as Tumblr. The ability of Tumblr users to wear their interests on their digital sleeves—and to share, follow, tag, and curate media based solely on their own interests—gives them a powerful way to break out of the typical top-down, one-way arrangement that exists in science communication and science journalism. Microblogs allow us to go beyond simply allowing audiences to comment and react and review on our turf. They are built on a mode of self-expression that makes the end user a critical piece of the communication process. The memes, subcultures, and styles that exist in these self-sculpted communities give us a new language into which we can translate important science, and a new audience with which to communicate. If we wish to reach these undiscovered and underserved science fans, we should embrace and become part of the fandom.

JOE HANSON, Ph.D., is the creator of the PBS Digital Studios YouTube series and website *It's Okay To Be Smart*. He has also written for *Scientific American, Nautilus,* and *Wired,* and his writing has been featured in *The Open Laboratory: The Best Science Writing Online.*

Joe is based in Austin, Texas. Find him at his website, http://www.itsokaytobesmart.com, or follow him on Twitter, @jtotheizzoe.

Notes

1. Stats obtained from Tumblr, 2013 (personal communication from Tumblr employee).

2. Fernando Alfonso, "The Real Origins of Tumblr," *Daily Dot*, May 23, 2013, http://www.dailydot.com/business/origin-tumblr-anarchaia-projectionist-david-karp.

3. Jason Kottke, "Tumblelogs," Kottke.org, October 19, 2005, http://www.kottke.org/05/10/tumblelogs.

4. Erick Schonfeld, "Why David Karp Started Tumblr: Blogs Don't Work for Most People," *TechCrunch*, February 21, 2011, http://techcrunch.com/2011/02/21/founder-stories-why-david-karp-started-tumblr-blogs-dont-work-for-most-people.

5. Rob Walker, "Can Tumblr's David Karp Embrace Ads without Selling Out?" *New York Times*, July 12, 2012, http://www.nytimes.com/2012/07/15/magazine/can-tumblrs-david-karp-embrace-ads-without-selling-out.html?pagewanted=all&_r=0.

6. Dan Zarrella, "How to Get More Likes and Comments on Tumblr," DanZarrella.com, http://danzarrella.com/infographic-how-to-get-more-likes-and-comments-on-tumblr.html#.

7. "In Conversation with Danielle Sterle," video, 15:30, from a presentation at Social Data Week on September 20, 2013, posted by Social Data Week, http://fora.tv/2013/09/20/In_Conversation_with_Danielle_Strle_of_Tumblr.

8. "Before the Federal Communications Commission in the Matter of Protecting and Promoting the Open Internet, Comments of Tumblr, Inc.," Federal Communications Commission, GN Docket 14–28, September 9, 2014, http://apps.fcc.gov/ecfs/comment/view?z=pk9t8&id=6018347452.

11

Science and the Art of Personal Storytelling

BEN LILLIE

Some of the best scientific stories are personal, using an author's experiences both to provide context for the scientific content and to create a relationship with the reader. Ben Lillie, co-founder of The Story Collider, *will describe how personal experience can help to inform the content of a blog, and how putting more of yourself into your writing can bring out the best in the story and the science.*

"Is there room for storytelling or personal experience in talking about science?" That's a question I hear a lot, and I find it deeply confusing. On the one hand, I get it: shaping science into a compelling story can distort the truth and, if it's done without care, distort it badly. But on the other hand, storytelling is the oldest art form; it's a way that people have been communicating and bonding with each other for millennia. The biggest cultural force in America—Hollywood—is obsessed with story and narrative. We grow up hearing stories and relating to them. When you sit down with your friends in the pub, you usually start by saying, "So, I was sitting on the bus when . . ." And you do that because that's how we relate. Yes, it takes care to do it right, but if you have something to say, why *wouldn't* you use story?

I've been obsessed with story for a very long time. From an early fascination with mythology to my career plan to be a playwright, it's been a constant part of my life, interrupted only briefly while I got a Ph.D. and did a postdoc in theoretical high-energy physics. Most recently I co-founded and continue to run *The Story Collider,* where we have people tell true, personal stories about science in their lives. The storytellers are people with science backgrounds and people without. The stories are all about their experiences, not about explaining the science. It's a fantastic way to show a side of science that rarely gets talked about. But the principles of narrative are far more generally applicable, and so are important for many kinds of science writing.

Straight "explainers," like this chapter, can be great, and are often the best choice for getting a point across, but I do think narratives and personal experience are vastly underused. That's understandable. Particularly if you're a scientist who blogs, it's easiest to default to a writing style you know, and science papers are (for very good reasons) written in the third person, without a heroic narrative arc. But stories can engage, they can pull people into a subject, they can be the thread that keeps them reading through the whole piece. The strongest stories can forge an emotional bond that helps people connect to information—and each other—in ways that straight facts usually don't.

Now, narrative theory is a huge field. Even the simplest question,

"What is a story?," turns out not to have any sort of simple answer. Without going too far down the rabbit hole, I'll say that there are a lot of definitions, none of which are great but some of which are very useful. For this chapter, I'll focus on two: one where the science is the main focus, and the personal experience is used as an aid, and the other where the personal story is the main point.

Science First

Steven Jay Gould had a trick, a very good trick, that he used all the time. He'd introduce an essay with a little personal anecdote, describing himself, say, walking on the beach looking for snails or watching a baseball game. He would then relate that narrative to a bit of science he wanted to discuss, for instance, ecological diversity within species or statistical analysis. But crucially, the anecdote often wasn't directly about the science he was discussing. He might use an example of learning baseball statistics to help illustrate how to interpret statistics of dinosaur fossils near the Cretaceous extinction. Or famously, he opened his book *Full House* with his own cancer diagnosis as a way into the arcana of statistical analysis. It wasn't necessarily "I was in the lab and we discovered this, let me tell you about it," but rather, "Here's a thing, it's kind of like this other thing."

More generally, using personal anecdotes can help our understanding of a possibly unrelated bit of science. This is the first sense of story, which is a tale of a thing that happened. (Did I tell you about the time I smashed my face in trying to skateboard? Oh, man.) This is the kind of story you tell a bunch of friends at the bar. It could be a funny thing to break up a long bit of explanation, a bit of connective tissue between one idea and the next. Or, as in Gould's case, it could be a useful metaphor. In any of those cases, it's incredibly useful for grounding the science, which can seem quite arcane compared to daily experience.

The other day I was trying to explain to a friend what it looks like inside a proton. The inside of a proton can be hard to visualize if you aren't used to it, to say the least. So I told him about the time I put a bunch of food coloring in a pot of boiling water, how the colors bubbled and popped and then mixed into a brown goop. That's sort of what happens to the quarks and gluons inside a proton. Now, that's not at all what it looks like inside a proton (it doesn't look like anything), but it's a good place to start building a metaphor. Putting it in the first person grounds it in the real world. Putting food coloring in a pot is kind of weird in itself, but saying that I did it for kicks helps people believe in it as a thing, and relate a little bit. (Who hasn't done a weird experiment in the kitchen?)

The challenge here is to make sure that the bits of story are serving the science that you want to discuss. If the story is for a metaphor, is that metaphor clear? If it's a funny moment to break up some explanation, is it actually funny? The key to making sure bits serve their purpose, generally, is to keep them short and leave out any details you don't need.

Story First

The other way to use your personal experience, and the one I'm much more familiar with, is to make the story the whole point of the piece. This is the kind of story that takes your readers on a journey with you, and brings them to the end having seen you change. This is a story in the sense that it's used in Hollywood, or in novels, or in the kind of live storytelling that *The Story Collider* is a part of. In its simplest formulation this kind of story is something with a beginning, middle, and end. In between, someone changes. That is, there's a plot and an "arc"—something about a character is different than it was when the story started. Maybe a relationship changed, or maybe the change is in how one sees the world, or oneself. The plot is what keeps readers going—they want to know what happens next—whereas the arc is why it's satisfying at the end.

These pieces accomplish a very different goal. Rather than teaching a lot of science, they help forge an emotional connection. It's not about getting readers to learn something, but about showing them why they might care. If all goes well, readers identify with the protagonist and care about the things the protagonist cares about. If you are the protagonist in a story about trying and failing to get data about the distribution of dark matter in the universe, and you tell the story well, in the end we'll care about the distribution of dark matter. We might not *understand* about the distribution of dark matter, at least not at any level of detail, but for this sort of story that isn't the point. Ideally once people care, they'll go looking for more information. Keeping that distinction in mind is essential, because of the next point: for this strategy to succeed you need to do the exact opposite of what I described for the previous technique—that is, the science bits have to be kept to an absolute minimum, and everything that survives has to be in the service of the story.

That can be really hard to do, and often will mean being okay with your audience not fully understanding something. I call this the Dilithium Crystal Principle. It's okay to have something in your story that isn't explained, as long as it's clear that it's important. In *Star Trek,* the characters need dilithium crystals to make the ship go. Why? How? No clue. But it's often an important plot point, and that's okay. They need to align the crystals, often as a matter of extreme urgency, and if they stopped to explain why, we'd lose the momentum of the episode.

One of my favorite examples of this is Tom Stoppard's *Arcadia,* which is quite possibly the finest piece of science writing ever produced. A couple of the characters study chaos theory and thermodynamics. And while the audience won't learn anything beyond "disorder increases" and "it's impossible to predict some things," those themes infuse the play so thoroughly it's impossible to come away without a changed appreciation of both.

So with all that in mind, here are some additional suggestions that we often give out to writers who are hoping to improve their storytelling.

Begin in the middle. Start with action. That doesn't necessarily mean a fistfight, but start at an important moment. Let's say you're telling a story about getting a wrong result and what it took to fix the problem. (A good sort of story!) Start with the wrong discovery. "I'm sitting in my lab, and I see the crystals come together, and I yell, 'Rachel, we found it!'" That's much stronger than "We were trying to figure out what to study, and figured crystal structure might be interesting, so we set up an experiment." All of that is implied in the fact that you're looking at crystals in a lab, and if there are bits you need to expand on (why did you think it might be interesting?) you can fill them in later. The important thing is to get the narrative momentum going from the very first sentence.

End at the end. If you have ever been at a live storytelling event you will see a remarkable thing: storytellers continuing on after the end of their story in order explain what it means. But whenever this happens, you will also see the audience become bored: they will start shifting in their seats, whispering to each other, or getting up to use the restroom. If you're pulling the audience along with a plot, once it's over, they're done. If you really want to explain what everything means (and most of the time, honestly, don't), then find a way to do it before you resolve the action.

Don't give away the end. Or any information really. There's a tendency to want to let readers know where the story is going, particularly if you want to be sure not to upset them. "Don't worry, we made it, but . . ." Don't do that. This is part of what will make the audience want to stick around. They'll want to know the answer to the key question: what happens next?

Set things up ahead of time. The flip side of not giving away too much is that if there's information that's essential to a plot point, set it up ahead of time. Let's say you're working with acetone, and you accidentally hold a match over it and it explodes because it's flammable. Tell us that it's flammable long before then, so we know what's happening the instant the match hits the liquid.

Know what the narrative arc is. You might not say it outright in the story, but know what changed from the beginning to the end about your character. Maybe you realized your dad is actually a smart person, or that your teacher really did care about you, or that you weren't alone in the universe. If you know what that arc is, it'll help you figure out what to include and what to leave out.

Make sure we know how you're feeling. One of the easiest ways to go wrong with writing about science is to forget that a lot of your audience doesn't find it exciting to watch hexagonal crystals form. So tell them. That might mean explaining the science, but you can also do it by simply telling us that you find them incredibly exciting. Early in the story set up that you love looking for hexagons, then when we get to them in the story ("Wow! I nearly jumped out of my seat—there they were, perfect hexagons!"), we'll trust you. Technical bits are exciting as long as you show the emotion.

Remember that storytelling is the kind of field you can spend your life learning about, with many layers. Every one of these "rules" can be broken, if you know what you're doing. If you want to read more about creating narratives, I'd recommend books on screenwriting. Film is a place where people obsess about how to make good stories. Many people like *Story,* by Robert McKee. Blake Snyder's *Save the Cat* is shorter and more direct. There are also the classic fiction-writing books *The Seven Basic Plots* by Christopher Booker, and *On Writing* by Stephen King. For live storytelling there isn't really a good book, although Margot Leitman has one coming out soon, and William Demastes's *Spalding Gray's America* has a lot of good advice, although it's mainly a biography of Gray. For the style where the story is anecdotes, most of Steven Jay Gould's work, particularly the mid to later essay collections, provides great examples (http://www.stephenjaygould.org). I'd also recommend anything by Lewis Thomas or Loren Eiseley, as well as Janna Levin's *How the Universe Got Its Spots.*

As to the opening question of whether one should use narrative, particularly personal stories, in writing about science, I can offer this: narrative is incredibly difficult to get right, and fraught with dangers of oversimplification. But it's also how we, as human beings, relate best to each other. It's how we bond with our friends and family and, when everything goes completely right, our enemies. Saying "You know, this happened to me" is the act of creating an anecdote and, as we all know, does not convey data. It's also the most powerful tool for communication in existence, driving all the profits of Hollywood and all the literature from centuries and millennia past that we remember today. Science writers ignore it at the risk of never being heard.

BEN LILLIE is the founder and director of *The Story Collider,* where people tell stories about their personal experiences with science. He is also a *Moth StorySLAM* champion and a former writer for TED.com.

Ben is based in New York City. Find him at his Tumblr, http://tumblr.benlillie.com, or follow him on Twitter, @benlillie.

12

Using Social Media to Diversify Science

ALBERTO I. ROCA

The discrimination faced in the workplace by minorities is just as prevalent on the Internet. But at the same time, social media can be an excellent resource for minorities in science communication, giving a boost to voices that need to be heard and increasing access to mentorship and role models. Alberto Roca, developer of minoritypostdoc.org, will explain how minorities can find a bigger voice through blogging.

Life is not fair. Or just. Or equitable. Competition is the norm in life, and a science career is no different as one strives for resources, data, accomplishments, and recognition. It's a lack of recognition that will strike a young trainee the hardest when the exhilaration of a new discovery subsides. A novel result must be communicated or one risks that the knowledge will be buried or (worse) disseminated by someone else. The common mantra "publish or perish" encapsulates this view, though I push the clarion call further as "communicate to consummate, else capitulate." Do not let your science results lie dormant in the science technical literature. That's what you risk when your work has no public exposure.

Broad communication via social media can bring attention to science and, more importantly, to the scientist. In the competition to establish and maintain a productive science career, recognition leads to more resources and opportunities. Unfortunately in this contest, the challenges for minorities are magnified either due to power imbalances or, by definition, because the majority outnumbers the minority. Here I provide examples of how social media can be used to diversify the sciences for the benefit of underrepresented minorities in STEM fields, such as African Americans, Hispanics, and Native Americans in the life sciences; women in computer science; and (South) Asian Americans in engineering leadership positions.

Supporting Minorities in STEM Careers

Fifty years after the Civil Rights Act was signed in the United States, the nation is still grappling with structural inequities suffered by ethnic minorities.[1] With respect to STEM workforce demographics, the representation of African Americans, Hispanics, and Native Americans has not reached the general U.S. population levels of 13 percent, 17 percent, and 1.4 percent, respectively.[2] Notably, by 2050 the United States will be a majority minority nation, with the Hispanic/Latino population alone reaching almost 30 percent.[3] How do we encourage this growing minority population to be engaged in STEM discussions as informed citizens? One solution is for the nation to have more minority STEM role models. Social media can amplify the visibility and voice of such individuals.

I curate a Diversity Bloggers web page (http://www.minoritypost

doc.org/view/bloggers.html) that serves as a useful starting point for finding STEM minority scholars who post online. Some examples are scientist Danielle Lee, technologist Adria Richards, engineer Micella DeWhyse (a pseudonym), and mathematician Ron Buckmire.[4] All are African American, reflecting the fact that many of the STEM bloggers of color I have found are from the black community. While much smaller in number, minority bloggers from other populations include Dr. Isis (a pseudonym), who is Hispanic/Latino; Cynthia Coleman, a Native American; Jeremy Yoder, who identifies as part of the LGBT community; and Viet Le, an Asian American.[5] Minorities who want to learn how to blog and tweet about their science and culture could look to these bloggers as role models.

A sampling of posts reveals what these bloggers advocate. Lee is very vocal about holding black media publications accountable for accurate science reporting.[6] Dr. Isis has written about gender discrimination against students and has advised white men on how to discuss diversity.[7] Coleman is a communications professor studying how science is communicated in mass media channels; she has a special interest in issues that engage indigenous peoples, such as the Native American Graves Protection and Repatriation Act.[8]

Blog carnivals, which allow more scientists to contribute to online writing, can also be a resource for new bloggers. A blog carnival is a collection of blog posts by different authors but all on the same theme. A hosting editor will introduce the carnival theme on his or her own blog and then summarize the contributed posts, producing a mini-anthology. In 2009, while still a graduate student, Lee began the Diversity in Science Blogs Carnival as a way to celebrate past and current minority scientists with early editions tied to the celebratory events of Black History, Women's History, and Hispanic Heritage months.[9] Nine editions were published before a hiatus while Lee completed her thesis work. In 2011, I helped Dr. Lee bring back the carnival for another ten editions, all of which are archived on http://minoritypostdoc.org/view/bloggers.html#carnival. The reboot included new editions that highlight underrepresented populations who did not receive attention in the first series such as the LGBT, Native American, and Asian American communities.[10] The Diversity in Science carnival topics were not limited to biographies of minority individuals. For example, editions were also published on issues such as the imposter syndrome and environmental awareness.[11]

The Diversity in Science Blog Carnival stories describe what minority students and professionals can face during a STEM career.[12] The following kinds of stories can seed your own writing about STEM minority communities:

- inspiring personal or career stories about science leaders;
- historical narratives about science, scientists, and community dynamics;
- reflections about one's own identity and its influence on one's career or science;
- stories on coming to terms with being a minority in a majority environment;
- explorations of subpopulation identity issues in fields such as science and medicine;
- helpful descriptions of career and professional resources (such as websites, articles, books, events, funding opportunities);
- information on outreach and mentoring activities that allow one to give back to the community;
- advocacy stories and leadership opportunities;
- descriptions of work policies that promote an inclusive, "safe space" environment;
- reflections on being a minority within a minority such as a woman scientist of color;

- information on educating and building relations with allies;
- stories uniquely beneficial to a specific, underserved sub-population, like gender identity in the LGBT community.

Becoming an Ally

In a workplace context, diversity refers to individual differences relative to the majority (or power-wielding) group where, in this U.S.-centric STEM discussion, the dominant group is composed of white, heterosexual, able-bodied, middle-to-upper-class males. The differences typically relate to an individual's identity regarding race, ethnicity, disability, socioeconomic status, religion, sexual orientation, gender, veteran status, age, national origin, or other personal characteristics. In an educational institute, workplace, or community, the dominant group establishes standards of practice that affect a minority person's perception of acceptance and inclusion. How welcoming does the environment feel?

Allies from the dominant group who are concerned about their community's minority representation can support interventions that educate, recruit, and retain underrepresented individuals in a STEM career. A dominant group's privilege would be extended (or checked) to allow minorities the kind of access to the community that had previously been impeded due to reasons such as historical injustices, existing discrimination, and other barriers. If a majority individual does not recognize his or her privilege as exclusionary, then this unconscious bias may prevent diversity efforts.

In the United States, STEM diversity interventions are commonly justified by two imperatives. First, there is a moral justification for initiatives that attempt to ameliorate past injustices to ethnic populations. The second imperative is a competitiveness argument that a nation's productivity and sustainability depend on the full participation of its population, especially in STEM fields that contribute to today's knowledge-based economy.

How can allies promote the careers and accomplishments of underrepresented minorities in the sciences? Attention could be drawn to diversity issues (interventions, advocacy, and so on) as well as to the minority scholars themselves as exemplified by the Diversity in Science Blog Carnival story ideas mentioned earlier. Let's examine the case of the diversity action plans that are expected to be in place at institutions receiving federal funds from the National Institutes of Health (NIH) and the National Science Foundation. For example, the NIH has inclusion-policy guidelines for scientific meetings as well as institutional research training grants.[13] Blogging and social media can be used to remind the greater community about such goals. For example, Jonathan Eisen advocates for greater diversity among conference speakers.[14] He calls out science conferences that have a low number of female speakers (#YAMMM means Yet Another Mostly Male Meeting), and boycotts invitations if the meeting organizers do not make adjustments to the speaker roster.

Another ally is engineer Suzanne Franks, who has been writing since 2006 at her blog *Thus Spake Zuska*. She describes herself as a "Goddess of Science, Empress of Engineering, and Avenging Angel of Angry Women" whose posts "offer the web's most excellent and informative rants on the intransigent refusal of engineering and science to open their doors to anyone but white males." Franks has long written about race and ethnicity, starting with a call to action about the lack of diversity in science blogs.[15] Back in the early days of science blogging (the mid-2000s), her comment discussions were quite lively, especially when, for example, she drew attention to racist comments by James Watson about African people.[16] Of course, the "ally" label is fluid because being a "minority" depends on the context of your environment or the topic discussed.[17] Thus

Franks can also speak personally as a "minority" when describing her working-class background and advocacy for the disability community.[18] This ally-minority duality is what motivated me to adopt the personal motto "We are all minorities, so let's help each other."[19]

More generally, allies can question the lack of diversity within their own discipline, as was done by marine biologist Miriam Goldstein.[20] The resulting discussion across more than seventy comments over six months included personal testimonials and intervention strategies by others. For example, Goldstein and the commenters noted that some expectations in the field sciences can impose barriers that not all students can overcome. Needing to have a car to reach a field site or participating in an unpaid summer research experience might mean that only students with the financial means for those "luxuries" will apply. Importantly, Goldstein's post motivated Lee to publish her own diversity manifesto about why young minorities are discouraged from STEM careers.[21] Lee describes three reasons why kids, especially from inner-city or working-class families, are discouraged from science: (1) lack of resources, (2) benign discouragement by well-meaning adults, and (3) active exclusion by powerful gatekeepers.

These messages about diversity by individual allies have been amplified by social media, and the resulting national discussions would have been impossible before the Internet age. Previously such discourse was the purview of local institutional gatherings convened by leaders in academia, funding agencies, policy think tanks, and other stakeholders of diversity. Social media has now democratized the ability of minorities and their allies to contribute to these policy discussions.

Allies can learn about minority populations from the advocacy and mentoring work of organizations that I call "diversity stakeholders." I publish a roster of diversity-aware higher education professional societies and conferences that is alphabetically arranged by cultural identity for easy browsing (http://www.minoritypostdoc.org/view/stakeholders.html). Over sixty nonprofit membership societies and at least twenty conferences serve as a critical mass of individuals from each of these respective communities. Many are stratified by scholarly discipline, such as the National Society of Black Engineers (nsbe.org), which has over eight thousand attendees at its annual conference and so may well be the largest. I am most familiar with the Society for the Advancement of Chicanos and Native Americans in Science (sacnas.org), having been an active volunteer since 2003, when I launched the creation of activities for our postdoctoral cohort.[22] The oldest of these stakeholder organizations were created in the 1970s during an "academic" civil rights movement. They created friendly communities where minorities could discuss their scholarship while celebrating their culture. Their annual conferences are a safe space where a minority scholar can let his or her hair down without being judged. For example, SACNAS members who are in the minority at their home academic institutions, and so may be self-conscious about speaking in Spanish around their academic colleagues who are not bilingual, can enjoy spirited discussions in their native language.

If you are interested in equity and inclusion in the STEM disciplines, I encourage you to participate in one of these diversity conferences at least once in your career. You can practice your ally skills by listening attentively and by withholding any tendencies to control a conversation. Also, you can earn credibility by mentoring students and postdocs—for example, by providing constructive criticism during a poster session. Becoming a temporary minority within a cultural group that is not your own will open your eyes to being an "other."

As a sensitized ally, your blogging will benefit in two ways. First, networking with so many different students and professionals will allow you to diversify your sources when you need topics or interviews for future posts. Second, drawing attention to those organizations that promote diversity within the STEM fields will magnify their opportunities and accomplishments for the benefit of all. Unfortunately, most of these organizations are small, volunteer-driven operations that are taxed to their limit, especially in producing their annual conferences. These organizations' communication channels are usually limited to just a website and email announcements; they typically lack a vibrant social media strategy. By amplifying the achievements of these organizations, allies can help them in the competition for recognition.

Minority scientists can accelerate their careers through blogging: keeping a blog can help them to raise awareness about their scholarship as well as about the social justice issues they find most compelling. In the absence of such communication, their activities might have only a local effect rather than potentially national or international influence. Publicizing the work of diversity stakeholders will also help minority trainees and professionals find support systems for career success. Through these actions, social media can help us prevail against the current inequities prevalent in STEM fields.

ALBERTO ROCA is executive director of the nonprofit Diverse-Scholar. He is also founding editor of the web portal minoritypost doc.org, founder of the postdoc committee of the Society for the Advancement of Chicanos and Native Americans in Science, and co-founder of the diversity committee of the National Postdoctoral Association.

Alberto is based in Southern California. Find him on his site, http://www.minoritypostdoc.org, or follow him on Twitter, @Minority Postdoc.

Notes

1. Braden Goyette and Alissa Scheller, "15 Charts that Prove We're Far from Post-Racial," Black Voices, *Huffington Post,* July 2, 2014, http://www.huffingtonpost.com/2014/07/02/civil-rights-act-anniversary-racism-charts_n_5521104.html.

2. "State and County Quick Facts," U.S. Census Bureau, accessed February 15, 2015, http://quickfacts.census.gov/qfd/states/00000.html.

3. Jeffery S. Passel and D'vera Cohn, "U.S. Population Projections: 2005–2050," Pew Research Center, February 11, 2008, http://www.pewhispanic.org/2008/02/11/us-population-projections-2005-2050.

4. Danielle N. Lee, "The Urban Scientist," *Scientific American,* http://blogs.scientificamerican.com/urban-scientist; Adria Richards, "But You're a Girl," http://butyoureagirl.com; Micella Phoenix DeWhyse, "Educated Woman," *AAAS Career Magazine,* American Association for the Advancement of Science, http://sciencecareers.sciencemag.org/career_magazine/previous_issues/articles/2004_08_27/nodoi.10925311399901543611: Ron Buckmire, *The Mad Professah Lectures,* http://buckmire.blogspot.com/search/label/mathematics.

5. Dr Isis, accessed February 15, 2015, http://isisthescientist.com; Cynthia Coleman, "Musings on Native Science," accessed February 15, 2015, http://nativescience.wordpress.com; Jeremy B. Yoder, "Denim and Tweed," accessed February 15, 2015, http://denimandtweed.jbyoder.org/category/queer; Viet Le, "Amasian Science," February 15, 2015, http://amasianv.wordpress.com/author/amasianv.

6. Danielle N. Lee, "Getting the Science Right in the Black Press—Making Headway with @EbonyMag," *The Urban Scientist* (blog), *Scientific American,* February 6, 2013, http://blogs.scientificamerican.com/urban-scientist/2013/02/06/getting-the-science-right-in-the-black-press-making-headway-with-ebonymag.

7. Dr Isis, "Shameful Gender Discrimination at UC Davis Veterinary School," *IsisTheScientist,* January 12, 2011, http://isisthescientist.com/2011/01/12/that_b_on_your_transcript_is_f; Dr Isis, "The Straight, White Dudes' Guide to Discussing Diversity," *LadyBits,* July 5, 2013, https://medium.com/ladybits-on-medium/53aaf639fc0c.

8. Cynthia Coleman, "Return the Bones," *Musings on Native Science,* March 3, 2014, http://nativescience.wordpress.com/2014/03/03/return-the-bones.

9. Danielle N. Lee, "Diversity in Science #1: Black History Month Celebration," *Urban Science Adventures!,* February 24, 2009, http://urbanscience.blogspot.com/2009/02/diversity-in-science-1-black-history.html; Suzanne E. Franks, "Diversity in Science Carnival: Women Achievers in STEM—Past and Present," *Thus Spake Zuska,* Scientopia, March 26, 2009, http://scientopia.org/blogs/thusspakezuska/2009/03/26/diversity-in-science-carnival-women-achievers-in-stem-past-and-present; DrugMonkey, "Diversity in Science Carnival #3: Celebrating Hispanic Heritage Month," *DrugMonkey,* Scientopia, October 16, 2009, http://scientopia.org/

blogs/drugmonkey/2009/10/16/diversity-in-science-carnival-3-celebrating-hispanic-heritage-month.

10. Jeremy Yoder, "Diversity in Science Carnival: Pride Month 2011," *Denim and Tweed,* June 30, 2011, http://denimandtweed.jbyoder.org/2011/06/diversity-in-science-carnival-pride-month-2011; Danielle N. Lee, "Diversity in Science Carnival #11: Native American Heritage Month," *The Urban Scientist* (blog), *Scientific American,* November 28, 2011, http://blogs.scientificamerican.com/urban-scientist/2011/11/28/diversity-in-science-carnival-native-american-heritage-month-2; Sabrina Bonaparte, "Diversity in Science Carnival #16: Asian-Pacific Heritage Month," *UW SACNAS Student Chapter Blog,* May 31, 2012, http://uwsacnas.wordpress.com/2012/05/31/diversity-in-science-carnival-16-asian-pacific-heritage-month.

11. Bethany Brookshire, "Diversity in Science Carnival: Imposter Syndrome Edition!," *Neurotic Physiology,* Scientopia, April 30, 2012, http://scientopia.org/blogs/scicurious/2012/04/30/diversity-in-science-carnival-imposter-syndrome-edition; Dianne Glave, "A Scratch-n-Sniff 'All Shades of Green' Blog Carnival," *Rooted in the Earth,* April 27, 2010: http://dianneglave.wordpress.com/2010/04/27/a-scratch-n-sniff-all-shades-of-green-blog-carnival.

12. Alberto I. Roca and Jeremy B. Yoder, "Online LGBT Pride: Diversity in Science Blog Carnival," *Minority Postdoc,* September 10, 2011, http://www.minoritypostdoc.org/view/2011-2-1-roca-carnival.html.

13. "Guidelines for Inclusion of Women, Minorities, and Persons with Disabilities in NIH-Supported Conference Grants," National Institutes of Health, accessed February 15, 2015, http://grants1.nih.gov/grants/guide/notice-files/NOT-OD-03-066.html; "What Groups Does NIH Consider to Be in Need of a Special Recruitment and Retention Plan in Order to Diversify the Biomedical, Behavioral, Clinical, and Social Sciences Workforce?" Frequently Asked Question: Recruitment and Retention Plan to Enhance Diversity, National Institutes of Health, last modified November 6, 2009, http://grants.nih.gov/training/faq_diversity.htm#867.

14. Jonathan A. Eisen, "Diversity (of Speakers, Participants) at Meetings: Do Something about It," *The Tree of Life,* May 29, 2012, http://phylogenomics.blogspot.com/2012/05/diversity-of-speakers-participants-at.html.

15. Suzanne E. Franks, "Archive for the 'Race Matters' Category," *Thus Spake Zuska,* Scientopia, accessed February 15, 2015, http://scientopia.org/blogs/thusspakezuska/category/race-matters; Suzanne E. Franks, "Where Are the Science and Race Blogs?," *Thus Spake Zuska,* Scientopia, October 20, 2006, http://scientopia.org/blogs/thusspakezuska/2006/10/20/where-are-the-science-race-blogs.

16. Suzanne E. Franks, "Watson to Africa: You're All Dumb," *Thus Spake Zuska,* Scientopia, October 17, 2007, http://scientopia.org/blogs/thusspakezuska/2007/10/17/watson-to-africa-youre-all-dumb.

17. Alberto I. Roca and Jeremy B. Yoder, "Online LGBT Pride: Diversity in Science

Blog Carnival," *Minority Postdoc,* September 10, 2011, http://www.minoritypostdoc.org/view/2011-2-1-roca-carnival.html.

18. Suzanne E. Franks, "Patch Hunky, PhD," *Thus Spake Zuska,* Scientopia, March 13, 2008, http://scientopia.org/blogs/thusspakezuska/2008/03/13/patch-hunky-phd; Suzanne E. Franks, "Archive for the 'Making Disability Visible' Category," *Thus Spake Zuska,* Scientopia, accessed February 15, 2015, http://scientopia.org/blogs/thusspakezuska/category/making-disability-visible.

19. Alberto I. Roca, "We Are All Minorities, so Let's Help Each Other: Introducing DiverseScholar," *Minority Postdoc,* November 9, 2001, http://www.minoritypostdoc.org/view/2011-2-0-roca-welcome.html.

20. Miriam Goldstein, "A Field Guide to Privilege in Marine Science: Some Reasons Why We Lack Diversity," *Deep Sea News,* January 24, 2013, http://deepseanewscom/2013/01/a-field-guide-to-privilege-in-marine-science-some-reasons-why-we-lack-diversity.

21. Danielle N. Lee, "A Dream Deferred: How Access to STEM Is Denied to Many Students before They Get in the Door Good," *The Urban Scientist, Scientific American,* January 24, 2013, http://blogs.scientificamerican.com/urban-scientist/2013/01/24/a-dream-deferred-how-access-to-stem-is-denied-to-many-students-before-they-get-in-the-door-good.

22. Alberto I. Roca, "20/20 Foresight—The New Postdoc Programs and LinkedIn Group of SACNAS," *Minority Postdoc,* October 4, 2010, http://www.minoritypostdoc.org/view/2005-roca-SACNAS.html.

13

I'm Not Going to Tell You How to Be a Woman Science Blogger

KATE CLANCY

Recent studies have highlighted the gender disparity that still exists in scientific fields, and women as science bloggers can face similar pressures. Kate Clancy, author of the blog Context and Variation, *gives valuable suggestions for how to combat sexism while advancing your own work.*

If you are reading this chapter, it is likely that either you identify as a woman, or you know someone who does. Good for you! We're pretty cool folk. There is also a good chance that you are thinking ahead to the kind of casual sexism, alienation, discrimination, or labeling that may happen to you as a result of the gender with which you or someone you care about identifies. Less good for you! But I understand why you are worried: we have all either experienced directly, or heard stories about, the unpleasantness that sometimes goes with being a woman on the Internet.

Is it harder to be taken seriously if you are a woman? Sure. In a cognitive study at the University of Nebraska–Lincoln, researchers found that both male and female research participants processed images of men globally, meaning that men could be perceived as a whole entity. Images of women, by contrast, were processed locally, meaning they were objectified as body parts.[1] There are too many scientific papers to count on differences in the perceived value of male and female writing—though there are several recent papers worth mentioning on differences in citation counts, and the positioning of male and female authors in academic papers.[2] Colleagues in anthropology have found that women have been underrepresented as speakers in a major primatology conference, and invited as symposia speakers far less often when men are the symposia organizers.[3] You can even take the same piece of writing—a résumé for a lab technician—slap a male or female name on it, and find that both male and female professors prefer to hire the man.[4]

Between these data and seemingly incessant cultural conditioning, women are taught their whole lives that they are vulnerable. This means that many of them feel the need to enter warily into every new situation. They can't enjoy a night stroll because they have to watch for men in the bushes; they can't have a drink at the bar without making sure no one puts anything in it. In a similar vein, women in science often feel like they need to overperform to prove that they deserve to be where they are. Unfortunately, even when women try to overperform, awareness of gender stereotypes often leads them to feel that their underperformance is inevitable.[5] So in addition to being seen by some as less competent due only to our gender, our many years of this conditioning have made us feel less competent than our peers, which may in turn cause us to speak less competently.[6]

Stereotype threat is when your awareness of a negative stereotype triggers a subpar performance in an activity supposedly influenced by that stereotype.[7] This awareness can be triggered from the smallest of acts: a woman can be asked to check "male" or "female" in a demographics box before starting a math test, or a proctor can remind a group of students that men tend to outperform women in the task they are about to complete. Stereotype threat has been shown to affect girls' and women's performance in STEM, even among those with positive attitudes toward these fields.[8] Stereotype threat messes with your ability to hold multiple pieces of information in your head at one time, makes you stressed out, and can cause you to be vigilant.[9] Add to that the vigilance from stories about women who write online being stalked or threatened, and you have a fair bit of your brain occupied with being afraid.[10]

Here's another thing to keep in mind: women and men have been conditioned to understand praise differently. Preschool boys and girls have been found to be motivated by several types of praise. But by the time boys and girls get to fourth and fifth grade, boys are unaffected by the type of praise they receive. Girls, by contrast, are motivated by praise of the process of their effort, and of its end result, but they have been shown to be demotivated by praise about themselves as people.[11] Perhaps these results are consistent with how uncomfortable you get when someone praises you: it competes with the message, long transmitted to girls, that you are supposed to be humble.

It also competes with the message you may have learned that anything good that happens to you comes from luck or the support of others, whereas your failures are your own. In other words, women and men have learned to attribute successes and failures differently as well. Generally speaking, girls are more likely to develop attributional styles that lead to their crediting others for their successes and themselves for their failures. Boys, the lucky bastards, tend to do the opposite.

We need to stop. Stereotype threat and vigilance use up our creative energy. Not owning our successes keeps us uncertain of our worth. Together, these issues lead to impostor syndrome, procrastination, writer's block, and holding back from our best ideas. I know this because it has happened to me quite a bit over the years. I've had long bouts of writer's block worrying about which group of men would decide to attack my comments section that particular week, lingered for days on a single drive-by rape tweet from a stranger, and been slowed by internalized sexist bullying from women. That paralysis is normal and understandable. But it doesn't put content on the Internet.

We all deserve better lives than ones ruled by fear. And the world deserves our brightest, best selves.

Let's recalibrate by remembering all the reasons you considered becoming a science blogger in the first place. Here's the most obvious: like all humans, you are smart. I would also guess that you find science interesting, so interesting that you want to share it with others. You suspect—if you're being honest with yourself, you *know*—that you could be really good at it. And the idea that someone else could read and enjoy your writing, maybe several someones, gives you a thrill.

Now write down all of the ways that you are amazing on a sticky note—I know you won't be able to fit them all because you have only one of those little square ones, so just hit the high notes. Then put that sticky note in this book, on top of the first couple paragraphs of this chapter. (If you are reading this on an e-reader, put the sticky note on your desk or your laptop, or make a virtual sticky note for your desktop. Maybe get a tattoo on the inside of your forearm.)

The next time the tendrils of stereotype threat curl around your ankles, read that note, and they should withdraw. Give them a good stomp, and maybe also the finger, and they'll think twice about messing with you again.

So here's the thing. Now that you are being made aware of the possibility of those feelings manifesting, and the uselessness of always being vigilant for danger, I am going to tell you a little about some of the particulars of being a woman science blogger. As much as possible, it's important to try and absorb this information neutrally. It's just information, things you should know, but they don't need to rule you. If any of this starts to worry you or make you feel bad, look at your sticky note.

Breaking in to science blogging is easy—you just need to figure out fairly idiot-proof, free blogging apps. In the beginning, you'll find that no one particularly cares about you. The chances of any of your early work going viral or being widely shared are small. This means that many people who are underrepresented in science blogging can develop their voices with minimal trolling. But when you are ready, getting exposure might take some time. It might take more time, in fact, to have people retweet you, share your material, follow you on Facebook, or read you regularly, if you identify as female. It may be harder to get gigs, or you may be expected to cover a particular beat. Once you figure out one successful hook in your writing, it may be harder for people to see you as being able to develop many different hooks. Over the course of history, as well as in widely read stories ranging from folktales to modern fiction, white, straight men have been able to be all things. Every step a person takes away from the identity that culturally we see as fundamental narrows how we societally conceive of that individual.

You may deal with more trolls, and more challenges to your authority, if you are female and/or have other intersectional identities. Perhaps you know a topic really well; maybe you even have a Ph.D.

in it. You are still more likely to get called on your material, more likely to get questioned, frequently by those with less expertise than you. You may get harassed, stalked, or threatened. Some of these troubling comments may be by people you know, some by people just looking for someone to terrorize for a few seconds. If you express your gender in a way that is very normative, or non-normative, or if you have any other intersecting identities underrepresented in science, these things could happen to you more often.

Here is where I have to tread a very fine line, because this is where you expect me to tell you how to handle these things. For me to even attempt to do so in a general way presumes a certain universality of experience, and a certain amount of privilege. Advising you on how to deal with a reader who questions whether you got your math right would mean that I think that you just need better tools to deal with jerks. It would presume, too, that all of the men and women who make you feel bad are intentional in their behavior. But no number of tools—no amount of agency—is going to get at the dominant cultural paradigm that is uncomfortable with women who do science and math. So there are many ways a person can handle someone problematic—you can delete and ignore, you can mock, you can shame, you can answer seriously. The success you have at these different styles is highly dependent on your intersecting identities. If you are a woman of color, or young, or queer, or working class, or not able-bodied, you will have different experiences as you push against oppression.

Therefore, I want to think about the other side of agency, and that is institution. Institution is defined as both the broad cultural practices that set up the many assumptions we make about how the world is supposed to work, and the people who benefit from the setup being the way it is. Agency, by contrast, is the amount of power we ourselves have to enact change. So what are the policies governing appropriate conduct on Blogspot, WordPress, Twitter, and

Facebook? How about the ones that are put forth by the American Association for Advancement in Science or the National Association for Science Writers? What responsibilities do participants in these groups have, and how can you gather and organize like-minded people to make these policies better in terms of addressing unspoken social norms and cultural conditioning? All of this is important, meaningful work because it is all of our responsibility to make our lives better and, I'd like to think, our responsibility to leave this planet a little better than when we got here.

So what will happen to you if you openly identify as female on the Internet? Although some frustrating things may occur, here are just some of the positively awesome things I'm pretty confident will happen if you identify as female. You will very likely develop a posse. We women science bloggers stick together, so find us online. We will support you, and viciously attack your trolls, and empathize with your struggles, in our common desire to say delightful things about science. You will learn to express yourself about topics that you find interesting. You will hone your craft. You will try a few different things on your blog, and some will become regular features. You will win the admiration of others. Someone, probably several someones, will enjoy what you have written.

You will find role models. You'll meet your role models, and that will be a little bit of a letdown but mostly really cool. You will be a role model. Someone will read a science book because of you. Someone will major in science because of you. Someone will teach science to their children because of you. Someone will feel a little less alone because of you. Someone will find a new love because of you. The possibilities for what you can do as a science blogger who identifies as female transcend that identity. So read the rest of this book for the how, re-read that sticky note for why you're the one to do it, and write something.

KATE CLANCY is an assistant professor of anthropology at the University of Illinois Urbana-Champaign. She has conducted high-profile research that highlights some of the issues that women face in academic research environments. She has previously written for the *Scientific American Blog Network* and now blogs for her own site.

Kate is based in Urbana-Champaign, Illinois. Find her on her site at http://kateclancy.com, or follow her on Twitter, @KateClancy.

Notes

1. Sarah J. Gervais et al., "Seeing Women as Objects: The Sexual Body Part Recognition Bias," *European Journal of Social Psychology* 42, no. 6 (2012): 743–753.

2. Jevin D. West et al., "The Role of Gender in Scholarly Authorship," *PLoS ONE* 8, no. 7 (2013): e66212; Daniel Maliniak, Ryan M. Powers, and Barbara R. Walter, "The Gender Citation Gap in International Relations," *International Organization* 67, no. 4 (2013): 889–922; Sara McLaughlin Mitchell, Samantha Lange, and Holly Brus, "Gendered Citation Patterns in International Relations Journals," *International Studies Perspectives* 14, no. 4 (2013): 485–492.

3. Lynne A. Isbell, Truman P. Young, and Alexander H. Harcourt, "Stag Parties Linger: Continued Gender Bias in a Female-Rich Scientific Discipline," *PLoS ONE* 7, no. 11 (2012): e49682.

4. Corinee A. Moss-Racusin et al., "Science Faculty's Subtle Gender Biases Favor Male Students," *Proceedings of the National Academy of Sciences* 109, no. 41 (2012): 16474–16479.

5. Steven J. Spencer, Claude M. Steele, and Diane M. Quinn, "Stereotype Threat and Women's Math Performance," *Journal of Experimental Social Psychology* 35 (1999): 4–28.

6. Shannon E. Holleran et al., "Talking Shop and Shooting the Breeze: A Study of Workplace Conversation and Job Disengagement among STEM Faculty," *Social Psychological and Personality Science* 2, no. 1 (2011): 65–71.

7. Spencer, Steele, and Quinn, "Stereotype Threat and Women's Math Performance," 4–28.

8. Jenessa R. Shapiro and Amy M. Williams, "The Role of Stereotype Threats in Undermining Girls' and Women's Performance and Interest in STEM Fields," *Sex Roles* 66, no. 3 (2012): 175–183.

9. Claude M. Steele, "A Threat in the Air: How Stereotypes Shape Intellectual Identity and Performance," *American Psychologist* 5, no. 6 (1997): 613–629; B. Seibt and J. Förster, "Stereotype Threat and Performance: How Self-Stereotypes Influence

Processing by Inducing Regulatory Foci," *Journal of Personality and Social Psychology* 87, no. 1 (2004): 38–56.

10. Toni Schmader, Michael Johns, and Chad Forbes, "An Integrated Process Model of Stereotype Threat Effects on Performance," *Psychology Review* 115, no. 2 (2008): 336–356.

11. Jennifer Henderlong Corpus and Mark R. Lepper, "The Effects of Person versus Performance Praise on Children's Motivation: Gender and Age as Moderating Factors," *Educational Psychology* 27, no. 4 (2007): 487–450.

14

Blogging as an Early Career Journalist

COLIN SCHULTZ

Starting out as a journalist means building a network of connections, developing a reputation, and gaining the trust of your readership. Blogging has forged a new road to picking up and honing journalistic skills. Colin Schultz, blog editor at Hakai Magazine, *provides insight into how blogging can give a head start into the world of journalism.*

As a profession, journalism operates on the basis of one thing: trust. This fixation is easy to see: Walter Cronkite, the *CBS Evening News* anchor, was heralded as "the most trusted man in America." CNN bills itself as "The Most Trusted Name in News."

As an early career journalist, building trust goes hand-in-hand with building your reputation—for being accurate, interesting, original, fast, and on deadline. Consistently proving yourself makes editors want to open your emails, or take a chance on a pitch they may otherwise ignore. A solid reputation can convince hesitant sources to open up, and your reputation can help people trust your work even when the story you're writing seems unbelievable. People may or may not trust journalism, but if you have built a reputation for being honest, they will trust *you*.

For a budding journalist, blogging can be both a job and a way to build your reputation. The rise of blogging has created a new space for journalists to play in. It's also spawned a new career option, the "journalistic blogger"—a writer who uses the conventions, styles, and tools of the web, along with his or her journalistic skills and values, to tell new types of stories.

Blogging to Get a Job

For both the science-inclined journalist and the researcher-turned-writer, there's learning to be done all around.

When you're new to journalism you have no reputation—no real reason for anyone to believe you can do what you say you can do. This is true whether you're fresh from journalism school or you're a scientist leaving the bench.

Journalism isn't so much "a thing to be interested in" as a set of skills. Yet to be an effective journalist you need to have both—the curiosity and the passion to learn about the world, and the skills to turn that new knowledge into stories. A Ph.D. in hand may suggest you know your subject matter far better than most, but journalism is a different arena from academia, and it's one that requires a different set of tools—ones you've yet to prove. Can you find an interesting story? Can you conduct an interview? Can you write?

Building your reputation requires demonstrating that you've picked up the requisite skills. But how you practice them, and how you show them off, are up to you.

Journalism is one of those rare professions that one can break into without any formal training. There are a few well-worn paths to gaining experience and exposure, such as taking on an (often unpaid) internship or enrolling in a journalism program. Even so, talking someone into taking a chance on you—with no experience and no clips—is difficult.

Blogging has added a new way to develop and showcase your abilities. It's a way to build a portfolio of work without having to wait for anyone to let you do it. As a blogger you may often write for free, but compared to paying for school or having to move to a big city for an internship, it's a path worth considering.

It's also not necessarily an "either-or" decision so much as a "both-and" sort of choice. Blogging can complement other efforts to build your reputation and your journalistic portfolio.

Harnessing Serendipity

In his 2007 book *The Black Swan: The Impact of the Highly Improbable,* essayist and statistician Nassim Nicholas Taleb introduced the concept of "Black Swan events"—rare occurrences with outsized consequences that are impossible to see coming, but easy to explain after the fact.

Black Swan events shake up your life, either for good or for bad, and there's almost nothing you can do to force them to happen. All you can do, Taleb suggests, is be open to what he calls the "envelope of serendipity," and to be ready to leap when the opportunity strikes: "Remember that positive Black Swans have a necessary first step: you need to be exposed to them."

To Taleb this means moving to a big city and going to lots of parties—you never know whom you'll run into. For a budding journalist it also means being online: between blogging and social media, you never know who will run across your work.

When I was in journalism school I was working on an independent research project on science communication. As part of the project I conducted in-depth interviews with leading science journalists, trying to get their opinions on the state and direction of the field. The work could have easily rotted on my hard drive and in my adviser's inbox. But I had these great interviews, and I was transcribing them anyway, so I thought, why not throw them on my blog?

I wasn't really prepared for the attention I received. Soon my work was being shared and talked about not only by the journalists I'd interviewed, but also by others in the field. My project was highlighted by MIT's Knight Science Journalism Tracker and by Harvard University's Nieman Journalism Lab. And a post on the project by MIT's Charlie Petit caught the attention of the hiring manager at a position for which I was applying—leading to my first job as a science writer.

When you're new and you have no reputation to lose, the positive possibilities of exposing your work to as many eyeballs as you can far outweigh the risks. After all, writes Taleb in *The Black Swan:* "Many people do not realize that they are getting a lucky break in life when they get it. If a big publisher (or a big art dealer or a movie executive or a hotshot banker or a big thinker) suggests an appointment, cancel anything you have planned: you may never see such a window open up again."

Each path into journalism offers advantages. School gives formal training and a safe sphere in which to practice. An internship could theoretically transition into a job. But blogging gives you something different: freedom to play and experiment, to try things outside of the strictures of a course calendar or style guide—a chance to do something new, and to get noticed for it.

Blogging as a Job

At its core, blogging offers independence and freedom—the freedom to write how and what and when you want. To a certain extent, this autonomy extends even to professional blogging jobs.

A blog, of course, is a medium, not a style. But the way blogging tends to fit into the standard mainstream media ecosystem right now is that blogs are quick hits. They're the short, pithy pieces that writers use to respond to current events or highlight interesting nuggets. While much journalism is done in a hurry, blogging deadlines are measured in minutes or hours, not days, weeks, or months. And unlike other types of journalism, many of the support systems that journalists are used to working with—editors, copy editors, and fact-checkers—are gone.

As a blogger, ensuring the accuracy of what you're writing, and the quality of the way it is written, usually falls entirely on your shoulders. As your own reputation grows, or as you come to work for more prominent employers, so too grows the potential for damage when you invariably make mistakes.

Writing in a Hurry

The challenge when working as a paid blogger is to find ways to do interesting work that upholds your journalistic commitments while operating under incredible time pressure. There are various ways to tell interesting stories in a brief amount of time. With the benefit of a few years of experience working as a journalistic blogger, I've hit on a few strategies that seem to work for me. I've used these approaches to write about everything from art and politics to war, food, video games, and natural disasters. They are all particularly suitable to writing about science.

Tell the Story around the Story

Understandably, news writers put a premium on new things. But every event has context and background. Any important new study is built on previous work. For nearly every story you see in print, as obscure as the event may seem, a similar situation has probably played out at some point in history. Putting the news in historical or situational context is a way to enable people to not just read the news, but also understand it.

"Explainers" and "backgrounders," stories that describe how we got "here" from "there," are a powerful form of storytelling that is well suited to blogging. For examples of how well this can be done, look at outlets like the *Washington Post*'s *WonkBlog*, the *Atlantic*'s (now defunct) *The Wire*, or *Vox* to see the power of quick-turnaround context.

Combine Existing Facts in New Ways

Working as a journalistic blogger means trying to strike a balance between the competing demands of novelty, accuracy, and time. Working on short deadlines puts limitations on how much research you can do. But the accuracy of your stories is fundamental, and ensuring that is the top priority.

In that vein, another way to tell interesting, accurate stories is to combine existing facts from other people's already vetted work in new ways. Pulling a single thread out of a long feature here, adding some background from another story there—with a dash of color from your own experiences or a news peg from current events—is a great way to create something new from a handful of old parts.

Relying as a starting point on the work of other journalists or bloggers or institutions—those whom you know and trust to be accurate and who likely have more time for research and investigation than you do—can give you more confidence in the information underpinning your story. But beware the trap of reblogging, which is when a blogger takes someone else's story, frame, and facts, and simply repackages them. This is the kind of work that gives blogging a bad name. Not surprisingly, it also doesn't help your reputation as a journalist and doesn't produce stories that tend to do very well.

Trust Your Own Expertise

Everyone has passions and interests, things they know more about and keep tabs on. By trusting your own expertise or knowledge, you can find the stories that people less well-versed in your pet subject might have overlooked. This works especially well for writing about science—especially if you are a scientist-writer with a strong academic background. In just a fraction of the time it would take someone else to get brought up to speed, you can find, read, figure out, and write about a new piece of research.

This approach is a little bit slower, and a whole lot riskier, than the other two. Unlike writing about history or information already vetted by other journalists, generating stories in this way means that the onus of ensuring the story's accuracy falls even more firmly on your shoulders. But with care, skepticism, and a heavy use of caveats, it's possible to write something really new in a hurry.

Being a Useful Blogger

Blogging is more than just about commenting, snarking, and making lists of animated GIFs. It can be a very powerful medium. Because blogging's emphasis is on speed and because it offers the freedom to tackle stories in creative ways that fall outside of the norms of traditional journalism, it can give journalistic writers a way to make valuable contributions to the news ecosystem. By finding and highlighting important or interesting stories, explaining and contextualizing discoveries or events, or using their expertise to dig up stories on topics with a niche appeal, bloggers can tell stories that would often have been left untold.

Blogging is not a surefire way to break into journalism. There is no one way to begin a career in any field. Yet whereas school and internships often keep aspiring writers within tight constraints, blogging is all about freedom. It's a way to try new things, to showcase your skills, and to reap the rewards when you discover something that works. Blogging can be high risk and high reward, but those risks and rewards will be all yours.

COLIN SCHULTZ is an editor at *Hakai Magazine.* He has also written for *Smithsonian Magazine* and for the American Geophysical Union's newspaper, *Eos.*

Colin is based in Ontario. Find him at his website, https://colinschultz.wordpress.com, or follow him on Twitter, @_ColinS_.

15

Institutional Blogging

KARL BATES

Science blogging takes on a whole new dimension when it's done from inside an institution. Public information officers can bring a valuable perspective to the science blogosphere. Karl Bates, director of research information at Duke University, shows us how science blogging can be used to enhance the reputation of a university and its research.

An institutional public information officer's wildest, most wonderful dream fifteen years ago would have been the ability to tell campus stories in our own words, reach readers beyond the usual sphere of the campus community, and truly engage that audience in a conversation. Thanks to blogging, that's no longer a dream. It's real, it's affordable, and we can all do it.

Blogging, and social media, give research institutions a flexible, affordable way to share their scene and their science directly with the public. The more relaxed voice and the sense of improvisation that have become a part of blog culture have put an authentic, human face on researchers and the student experience. Students, faculty, administration, and staff can show some passion and personality and be conversational, genuine, and connected to readers.

If you're coming at this from a magazine or media relations perspective, it's important to recognize what's different about the blogging medium—it's not a lecture, it's not one-way. Anyone in the world can be part of your audience, and those who view blogs expect to be able to talk back and ask questions, whether through enabled comments or social media.

All institutional research blogs share a few basic needs. They have to be updated frequently to keep the attention and respect of both the readers and the search engines. The writing should be bright and engaging, not turgid and lecturish (hint—keeping posts short helps). And perhaps most important, every post should include something visual. Research institutions have some of the best eye candy anywhere on the Internet. Put it to use!

Staff Blogs

Most institutional science blogging is done by a news office staff or by communicators in the various departments: the folks who publicize an institution's research by writing press releases, newsletters, web content, and so on. A lot of these people are "recovering journalists" who have entered into blogging already knowing how to write a decent story.

Blogging gives these research communicators the remarkable new ability to speak directly to the public, rather than being stuck in the media relations rut. It's a cheap and easy way to update campus news. It also gives the news office a chance to publish stories about personalities and the scientific process that would never make the cut with reporters but are precisely the kinds of science stories we really need in order to explain to the public what we do.

Most organizations also use blogs as a handy way to publish smaller, more informal news that isn't worth an external press release or a spot in the campus newspaper or news site, like people winning awards or student accomplishments, small-but-cool findings, eye-candy images from an otherwise impenetrable paper, or events that would be of interest to only a subset of the campus community.

Although there are some very well-read blogs out there that feature magazine-length works, most blogs thrive on being both frequent and brief. It's good to aim for maybe three posts a week, each shorter than four hundred words (that's less than one page in Word), and to always have a photo of some sort. This high-frequency approach is more fun than creating longer clips less often, and it results in a much more compelling collection. Have more to say? Post more frequently. (Frequency, as we'll discuss a little later, builds readership.) Try to get a new photo in every few paragraphs. The collective experience of these posts, when searchable and carefully tagged, creates a mosaic of your campus's bigger research picture, much like a wall full of snapshots.

There are lots of ways to go about staffing an institutional research blog, but it's pretty clear that the institution and the bloggers have to commit to regular posts: sporadic and halfway just won't cut it. Some blogs are run by just one staffer who does it on his or her own, with uneven results. There are plenty of abandoned examples of these. Other blogs are formal, staff-driven publications that are integral to a larger communications strategy.

Representing what might be the outer limit of speed, copy volume, and staffing, the Stanford School of Medicine news office produces a massive and aptly named blog called *Scope* that publishes three to five times each day at about 250 words a pop. It has more than twenty regular authors and dozens more who contribute less frequently.

When it launched in 2008, *Scope* was aimed at a general audience, not just the media—which is still today a pretty radical notion for some news offices. "The blog gave us a way to fill a gap and promote our own work," says *Scope* editor Michelle Brandt, associate director of digital communications and media relations at the Stanford School of Medicine. But its staff also made the gutsy decision to cover medical news beyond Stanford's campus because "there weren't tons of high-quality, medical/science-focused blogs and we wanted to serve as a curated source of medical/research news for our audience," Brandt says. *Scope* does short coverage of research findings on and off campus; Q&A's on timely topics; thought pieces from faculty, students, and even patients; and expert reactions on major medical news. They can also use the blog to offer live coverage of an event.

Multimedia producers Jessica Wheelock and Zak Long in the University of California's Office of the President hit on the idea of using a Tumblr blog in early 2013 as a way to repurpose some of their video content and experiment with different ways of retelling research stories in general. The *U.C. Research* Tumblr (http://ucresearch.tumblr.com) currently has more than 130,000 followers, making it the largest social media product coming out of the Office of the President.

Tumblr is a more visual approach to blogging, which plays to one of the strengths of science news. And nearly half of the *U.C. Research* posts use an animated GIF, which makes the stories really jump off the page. The archive page showing months of colorful and intriguing scientific images dances around like Harry Potter's newspaper.

The copy with each image, by contrast, is little more than a long caption that refers readers to the original material, whether that's a news release on the UCLA news site, or animated photos of a science art installation done by a Cal Berkeley alumna. Tumblr hits a different community than traditional blogs or Facebook, attracts a strong audience for its science postings, and has an active culture of sharing and liking.

News officers around the University of California system can upload their content to the site or send it by email for Wheelock and Long to edit and post. That is, rather than have a group blog, they've chosen to keep editorial control central to maintain a consistent voice.

When the *U.C. Research* Tumblr is really cranking, it might carry two new posts a day, but on average the staff aims for one post a day, which takes about two hours of staff time every morning. In addition to curating visual research news that comes in from the news offices of the University of California system, Wheelock and Long also look at what's popular on Tumblr and elsewhere each day to see if there are any potential University of California ties they might feature. "We began on a trial run and once the followers began accumulating, it became easy to get people on board," Long says.

Even if you set your sights a little lower and stay within the confines of your campus, the blog can be a hungry beast. Try spicing things up with standing features like a mystery photo of the week or a throwback Thursday item from the archives. Heck, post a ton of photos and just write captions! Have a professor or grad student email serial updates from fieldwork in Antarctica or Kenya or Cleveland. Run a little contest of some sort.

There really are no rules, so you can experiment with different forms of coverage too. The *Vector Blog* at Boston Children's Hospital covered an on-campus conference with a neat little roundup: "Five Cool Medical Innovations We Saw Last Week," at just one paragraph and a small photo each. That's the blog equivalent of a tasty little bowl of M&Ms.

For big stories, like a discovery in physics that leads to submissions from 1,500 authors from hundreds of institutions, your blog can draw local interest if it focuses tightly on just your person's role in the bigger event. You might even find that other blogs and social media pick up these posts and give you more news credit than if you had tried to force a traditional mainstream pitch. (And while you're at it, try live-tweeting the discussion from a room where the people you're promoting are watching a remote event, like a teleconference from CERN or a Nobel Prize ceremony.)

Polishing Your Message

Do you need an editor? It's a good idea to have one if your blog is seen as an institutional product, however informal, coming from your .edu or .org domain. (If you're .gov, I figure you're getting reviewed ten times as it is.) Some other person ought to look at a post before it goes live, mostly just to catch embarrassing word deletions or spelling errors, or to make sure the headline is composed of six or seven words that will absolutely sparkle and reel in the eyeballs. The editor will also make sure the post is appropriate to the institution. Give your writers a long leash to maintain their sense of ownership and the authenticity of their voices. But although you want them to be real, sloppiness—or too much self-referential, first-person writing—will reflect poorly on the institution. (And avoid my pet peeve mistake —date your posts and note the author if it's a team blog.)

The rules are looser in blogs, but there still have to be some policies. Spend some time formulating expectations about appropriate language and content, and spell out any policies your campus might have about photography, unpublished data, and so on. Here's a good one: student bloggers should always inform their principal investigator, tour leader, host family, or other supervising group that they are blogging. No surprises.

What about source review? For formal press releases that are meant to be the account of record (and are likely to be picked up wholesale by other media), public information officers are expected to run copy by their sources before publication for fact-checking and comfort level. In the interest of timely posts, many research blogs have dispensed with this step. Still, to avoid unpleasant surprises, bloggers should approach a speaker either before or after an event to ask whether they may blog about it, and students should show their profile subjects what they've done. Because this is not quite the same level of fact-checking rigor that one would expect of an institutional press release, you should talk about the issue with others at your institution and decide what the ground rules should be in each kind of situation.

Administrator Blogs

Blogs from administrators are out there, and unfortunately most are about as compelling as that "message from the President/Dean/VP" you find just inside the cover of your campus magazines. Blogging is a great way to publish, and an even better way to engage an audience in conversation. But sometimes a blog isn't really the right tool for the content you want to get out there. Just because it's posted doesn't mean it will get an audience.

If one of your administrators asks for a blog, have a serious conversation about what a whole year of content would look like—how frequently would new material be posted, how long would the entries be, who's doing the work? Most importantly: who's the intended audience and is a blog really the right way to reach those folks? Are you trying to start a conversation and willing to sustain it? Will the content be something the audience wants to engage with and maybe even share with their networks? Then do a little legwork to find analogous blogs being produced elsewhere and ask their managers both what the traffic is like and what they have learned about successful content. It will be time well spent.

Blogging by Affiliates

In addition to staff bloggers, consider employing some work-study students as regular bloggers or inviting students, faculty, or staff to offer up one-time guest posts. Student bloggers gain by acquiring some really great content for their portfolios, and the news office benefits by having some very affordable eyes and ears in places that staff can't get to often enough. Undergraduate student bloggers can provide an authentic view of research on your campus that will be catnip for some of your prospective students. Work-study students cost only a couple of bucks an hour and will essentially be paid to do what an undergraduate ought to be doing anyway: attending brown bag talks, seeing visiting Nobel laureates, or acquiring a deep understanding of a fellow undergrad's research.

Another source of student copy is frustrated graduate students who want to see if this science writing thing could work for them. Because of the paperwork involved, you probably can't pay for this help, but a little editing and coaching can prove to be acceptable compensation in some cases. Perhaps some of these students will become regular contributors; a couple might even go on to become science writers.

Casual blog voice can be a struggle for some students, especially as they become Serious Scholars of Science in the upper classes. Successful students have unfortunately been rewarded by their professors for writing ever longer and wordier class assignments, and a blog requires just about the opposite style. So don't let them write term papers for you—insist on a conversational tone, lay vocabulary, and brevity. Then resist the temptation to overedit student blogger submissions; they're best when they're authentic.

Pictures are a must. Make sure that everybody who contributes to the blog understands what is and isn't fair game (steer them toward Wikimedia Commons and explain Creative Commons) or have them snap smart phone images while they're at an event or interview. Your bloggers should know the provenance of every image they post—Google image search can often turn up the real story.

Remember Your Audience

The one sure-fire way to ensure that your blog fails is to be infrequent and unreliable. "We've found that post frequency is really important," says Tom Ulrich of the staff-written *Vector Blog* at Boston Children's. When that blog's writers were called off on another project, frequency fell off to two posts a week and traffic fell with it. (Just imagine what the dean's three posts a year would be like!) Search engines use frequent updates as a proxy for knowing which blogs are engaging and active. If a human reader sees that you haven't posted in two months, she probably won't find it engaging or interesting either.

Think of your institutional blog as just one tool in the communications box. Like everything else you're producing, the blog's content should be repurposed into as many different channels as seem appropriate (campus news page, alumni email, admissions marketing, undergrad research office, Facebook, Instagram, and so on). Enlist the help and partnership of campus social media and publications and encourage them to feature your content. A Facebook post or a tweet will drive significant new readers to your blog. Make sure the campus newspaper folks are watching your feed, too—it can be another source of content for their hungry pages. You will see a definite traffic spike on the blog anytime a post is referenced somewhere else on campus.

Unless your institutional research blog is being used as your primary news outlet, however, it is probably not going to set the world on fire. So it's important to keep the cost-benefit ratio in mind and not get carried away: instead make sure the resources you put into the blog are proportional to your overarching communications goals. If, for example, you have cheap student labor and more than one staffer, it might be reasonable to aim for three to five posts per week during the academic year. If you're a solo practitioner spinning a lot of plates at once, three blog posts a week would probably eat up too much of your productivity.

A great post for the staff- and student-written *Duke Research Blog* is something that gets picked up on Twitter or arXive perhaps, and earns more than two thousand clicks. (The all-time champ as of this writing is 3,700.) The blog has about twelve thousand visitors per month, but a given post might garner only a few hundred clicks. Still, it's offering new and different kinds of research coverage, providing a training ground for students, and building "the long tail" of posts that search engines will pull up months or years down the road. And yes, we sometimes get media pickups from the blog.

No matter how great you thought your old research magazine may have been, there's just nothing like a blog for reaching new audiences and making them feel truly engaged. And compared with traditional print media, it's ridiculously cheap to start and operate a blog. I encourage you to give it a try.

KARL LEIF BATES is the director of research communications at Duke University. He's a former newspaper journalist and current public information officer.

Karl is based in Durham, N.C. Find him at research.duke.edu and the *Duke Research Blog,* sites.duke.edu/dukeresearch.

16

Blogging as a Resource for Science Education

MARIE-CLAIRE SHANAHAN

Online tools offer great potential for enriching science classrooms by bringing the processes and people of science to life. How can writers best target classroom audiences? How can educators leverage students' knowledge and excitement about online tools? Marie-Claire Shanahan is a blogger and associate professor of science education at the University of Calgary, where she investigates the ways that people communicate science online and in classrooms. Here she examines some of the key advantages of digital communications tools that truly engage students, offering general guidance for writers and creators as well as specific advice on how to develop online writing spaces where students are the authors.

There is a long history of bringing science news into classrooms as a way to generate interest and show students the frontiers of scientific knowledge. Many generations of science students will recall assignments asking them to cut out newspaper articles for a classroom bulletin board. Surveys show that almost all science teachers incorporate science news into their curricula in some way.[1] And students say they find science news interesting, motivating, and important.[2] Online writing spaces offer educators, scientists, writers, and students of all ages new ways to interact with each other and with the scientific community. Blogs, Twitter, Tumblr, and YouTube can provide flexible spaces for deeper and more contextual coverage of science news, facilitate interactions with experts outside of the classroom, and allow students to communicate as experts themselves.

The particular reasons that students enjoy science news in the classroom point to ways that digital tools can make valuable educational contributions. Students report that they find the novelty of science news fascinating; they like to feel that their knowledge is up-to-date. These days, when making school relevant to students has become an important rallying call, some students even say that the most relevant and useful topics are those that they can talk about with friends and family, such as breaking news stories and cutting-edge findings.[3] Online news sources, including blogs, have a terrific advantage in reporting and updating stories in almost real time. One good example was NASA's 2010 arsenic-based life study. Reported first in a live-streamed press conference, the paper was discussed simultaneously on Twitter and quickly appeared on science blogs and in online science news venues.[4] Students and teachers had access to the main research claims, and critiques, within moments of their being made public.

Students are attracted not just to the excitement of current science news; the form matters as well. One research team asked high school students to explain what they look for when they choose to read an article about science. They said they were more interested in pieces that were creative—that used metaphor, analogy, and poetic language to help explain concepts—than in ones written in a more formal, traditional journalistic tone.[5] This is an important observation for bloggers and other online writers. Online venues for science writing, especially blogs, are noted for the wider freedom that they give writers both for covering topics that are of personal interest, and for covering them in nontraditional ways.[6] There are fewer constraints in form and length, and online writers can use that to their advantage to draw in young readers. Randall Monroe, creator of *xkcd* comics, illustrates this well with his *What If* series (http://what-if-.xkcd.com), which uses informal language, humor, and absurd situations to explain complex science and math concepts.

Beyond generating interest, reading and understanding science news is one of the key skills associated with scientific literacy, especially when that literacy is taken to mean having more than just a basic understanding of scientific facts. Exposure to science news is very important for preparing students to be thoughtful decision makers on key socio-scientific issues such as climate change and biotechnology. And this is where conventional media, such as newspapers and television news, have typically come up short as a classroom resource: they tend to emphasize recent published findings and ignore the context of scientific discovery.[7] This contextual information is an important element of science curricula and standards across most English-speaking school systems and is essential for helping students make sense of scientific controversies and challenges. But short newspaper articles or television clips rarely have room for specific contextual details, such as how new findings support and contradict previous work, how funding decisions were made, how the peer-review process progressed, or other background information that helps illustrate the importance of argument and disagreement in science.

Online communications about science can move beyond these constraints. As previously mentioned, online sources were directly responsible for the timely reporting of research results related to the possibility of arsenic-based bacteria. More importantly, online sources (in particular Twitter and various types of blogs) were responsible for quickly making criticisms of the research public. A rich collection of different critiques was quickly available to students and teachers, providing an opportunity not only to learn about the downfalls of this individual study but also to develop a greater understanding of the processes of science. Science writer Carl Zimmer, for example, did not just publish an online piece reporting on the criticisms; he also published on his blog direct emails from thirteen scientists examining the experimental methods and findings in detail. Microbiologist Rosie Redfield wrote candidly on her concerns about the study and drafted a letter to the editor of *Science* with the help of commenters on her blog.[8] She also publicly documented how she attempted, and failed, to replicate some of the key experiments from the study.[9] These writings have been a tremendous resource for classroom teachers and have given students a way to read critiques straight from the fingertips of scientists (exasperated language and personal feelings included), so they can start to understand the role of those kinds of critiques in scientific research. Both of these examples illustrate how online writing spaces can provide materials and voices not often found in conventional science news. These contextual materials can help students to understand better the people and processes of science.

Another key problem that teachers and students encounter in reading conventional science news, especially about controversial topics, is that the credibility of the news is difficult to assess. Researchers and educators have argued for many years over how well students can be expected to assess the credibility of claims. When reading science news, teens are hesitant to use their background knowledge of a topic to determine the veracity of what they are reading. They tend to trust written texts too readily.[10] Others have noted that there are very few science news stories for which students, or anyone outside of the relevant field, can be expected to have the full background knowledge needed to assess the finer points of new research findings. For example, with the announcement of evidence for gravitational waves, many physicists had difficulty explaining, and even assessing, whether the complex findings represented a true breakthrough.[11] Stephan Norris, a science education researcher, has suggested that students should instead be taught to assess the credibility of the expert, such as their standing in the scientific community, their prior publications, and their place in any developing consensus. This seems sensible to some degree, but it is only in online spaces that it is even possible. Blogs and online news media that provide detailed networks of linked material, such as researchers' other work, their critics, and their own personal writing, are thus a potentially valuable resource for students learning to assess the credibility of scientific experts and the work they produce. This is more valuable than just access to a search engine because it helps to teach students where to look for this information and what types of sources are relevant and trustworthy—especially important skills when examining credibility as it relates to controversial or contentious scientific topics.

Online Spaces as Classroom Places

Online spaces such as blogs can be used in many ways by students and teachers themselves. For example, Staycle Duplichan, a Louisiana high school teacher, runs a classroom blog with student-created content. She has outlined eleven different types of posts that her students write for her classroom blogs.[12] These range from posts that start science book club discussions to those that share study tips among students.

She is not alone in the variety of writing and interaction tasks that she envisions for student blogs. Researchers followed nine classroom blogs created by middle and high school science teachers and written by their students. They noticed four formats that all of the blogs shared:

- scribe posts, where an individual student provides an account of the concepts addressed in class that day;
- resource-sharing posts, where many students contribute links to websites, simulations, videos, databases, and other resources relevant to the current class topics;
- opinion-solicitation posts, where the teacher asks each student to respond with his or her views on a controversial or contentious topic; and
- required responses, where the teacher or an invited guest writes a post and students are required to post comments in response.[13]

The idea of having students take turns acting as a scribe for the class is a particularly popular one among teachers. There is great value in students taking responsibility for communicating about what they learn, and it can be motivating to them to have an authentic and real audience for their writing, beyond their teacher.

There is also a strong connection between this type of blogging and the creation of explainer blog posts written by scientists and science writers. Joe Hanson, creator of the PBS video series *It's Okay to Be Smart,* sometimes writes posts to accompany his videos. These explainer posts fill in details and expand on concepts explained in the video just as students' scribe posts are meant to do.[14] The explainer posts illustrate the importance of online science writing not only as a resource for helping students understand science, but also as a model of good writing for students improving their own writing skills.

When given the opportunity to be active science bloggers, students seem to act in many of the same ways as those who blog outside of classroom settings. As April Luehmann and Jeremiah Frink noticed with the scribe posts, students will often dive into their assigned topics to create comprehensive and deep explanations. In a classroom science blog observed by other researchers, students also took on the self-corrective tasks sometimes ascribed to communities of science bloggers.[15] With their teacher, they created "the editor's initiative," whereby students earned credit for identifying errors and omissions in the daily scribe posts and for working with the authors, their peers, to correct the errors. This led to important discussions about writing standards and editorial practices in science.

Other online venues such as Twitter can also bring students in direct contact with scientists. A regular discussion using the hashtag #scistuchat is hosted on the platform by high school science teacher Adam Taylor. Every month, a new topic is presented and questions are posed to be answered by students and any scientists interested in participating. They are all encouraged to interact with each other directly and to share what they know and what they are curious about. In this space there is room to move beyond reporting on new findings to discuss the methods, connections, and contradictions that characterize the process of doing science. The hashtag is popular both among science teachers and among scientists, who are eager to share their passion for science with young students.

Science Blogs and the Classroom: Better Together

One of the main reasons that blogs and other online venues are increasingly seen as powerful tools for the classroom, in particular for science education, is the way that they can connect students to experts and communities outside of the classroom. Blogs can open doors to scientific work where scientists, writers, students, and any other interested individual can interact.[16] Blogs and other online venues provide the context behind science news, and their links allow students to assess the credibility of the scientists involved. Blogs also provide space for professional scientists to weigh in on their discoveries, as well as for discussions of scientific processes and the nature of science. There are opportunities for students to act as science writers and to learn from exemplars that they find online. Scientists, science writers, science educators, and students all benefit from the diversity of dynamic and richly interactive new platforms for sharing ideas online.

MARIE-CLAIRE SHANAHAN is an associate professor and research chair in science education and public engagement at the Werklund School of Education at the University of Calgary. She is a blogger at *Boundary Vision* and has written for the *Scientific American Blog Network, Story Collider Magazine, American Physical Society News,* and *Science and Children.*

Marie-Claire is based in Calgary, Canada. Find her at her website, http://boundaryvision.com, or follow her on Twitter, @mcshanahan.

Notes

1. Melissa R. Kachan, Sandra M. Guilbert, and Gay L. Bisanz, "Do Teachers Ask Students to Read News in Secondary Science? Evidence from the Canadian Context," *Science Education* 90, no. 3 (2006): 496–521.

2. Ruth Jarman and Billy McClune, *Developing Scientific Literacy: Using News Media in the Classroom* (New York: Open University Press, 2007).

3. Martina Nieswandt and Marie-Claire Shanahan, "'I Just Want the Credit!'—Perceived Instrumentality as the Main Characteristic of Boys' Motivation in a Grade 11 Science Course," *Research in Science Education* 38, no. 1 (2008): 3–29.

4. Olivier Dessibourg, "Arsenic-Based Bacteria Point to New Life Forms," *New Scientist*, December 2, 2010, http://www.newscientist.com/article/dn19805-arsenic based-bacteria-point-to-new-life-forms.html; Dennis Overbye, "Microbe Finds Arsenic Tasty; Redefines Life," *New York Times*, December 2, 2010, http://www.nytimes.com/2010/12/03/science/03arsenic.html.

5. Krystallia Halkia and Dimitris Mantzouridis, "Students' Views and Attitudes towards the Communication Code Used in Press Articles about Science," *International Journal of Science Education* 27, no. 12 (2005): 1395–1411.

6. Marie-Claire Shanahan, "Science Blogs as Boundary Layers: Creating and Understanding New Writer and Reader Interactions through Science Blogging," *Journalism* 12, no. 7 (2011): 903–919.

7. Billy McClune and Ruth Jarman, "Encouraging and Equipping Students to Engage Critically with Science in the News: What Can We Learn from the Literature?," *Studies in Science Education* 48, no. 1 (2012): 1–49.

8. Rosie Redfield, "We've Received the #arseniclife Reviews from Science," *RRResearch* (blog), March 16, 2012, http://rrresearch.fieldofscience.com/2012/03/weve-received-arseniclife-reviews-from.html.

9. Rosie Redfield, "How to Test the Arsenic-DNA Claims," *RRResearch* (blog), March 16, 2012, http://rrresearch.fieldofscience.com/2011/05/how-to-test-arsenic-dna-claims.html.

10. Stein D. Kolstø, "Scientific Literacy for Citizenship: Tools for Dealing with the Science Dimension of Controversial Socioscientific Issues," *Science Education* 85, no. 3 (2001): 291–310.

11. Phil Plait, "Cosmic News: Astronomers Find the Twisted Fingerprints of Inflation in the Background Glow of the Universe," *Slate*, March 17, 2014, http://www.slate.com/blogs/bad_astronomy/2014/03/17/evidence_of_inflation_astronomers_detect_gravitational_waves_from_the_early.html.

12. Staycle C. Duplichan, "Using Web Logs in the Science Classroom," *Science Scope* 33, no. 1 (2009): 33–37.

13. April Lynne Luehmann and Jeremiah Frink, "How Can Blogging Help Teachers Realize the Goals of Reform-Based Science Instruction? A Study of Nine Classroom Blogs," *Journal of Science Education and Technology* 18, no. 3 (2009): 275–290.

14. Joe Hanson, "The Cycle: Carbon and Oxygen and You," *It's Okay to Be Smart*, http://www.itsokaytobesmart.com/post/89005230922/this-post-is-an-explainer-to-go-along-with-this.

15. Robyn MacBride and April Lynn Luehmann, "Capitalizing on Emerging Technologies: A Case Study of Classroom Blogging," *School Science and Mathematics* 108, no. 5 (2008): 173–183.

16. Marie-Claire Shanahan, "Science Blogs as Boundary Layers: Creating and Understanding New Writer and Reader Interactions through Science Blogging," *Journalism* 12, no. 7 (2011): 903–919.

17

Communicating Science as a Graduate Student

JASON G. GOLDMAN

Many of the scientists blogging about science on the Internet are still in graduate school. By blogging while in graduate school, students may feel encouraged to explore new aspects of their research and may further develop their writing skills. Jason G. Goldman, who started his blog The Thoughtful Animal *while in graduate school, explores the pros and cons of writing as a graduate student, and suggests how you can use your blog to help your career as a young scientist.*

Life as a graduate student can be hard, but you probably don't need me to tell you that. That's true whether you're a graduate student in the sciences or in the humanities, or in a professional school like business or journalism. You've got to balance some combination of research, classwork, and teaching commitments. You have to get yourself to conferences in order to network and to present your posters and papers to your colleagues. If you're in a professional school, you probably have an internship or two to complete. You need to write and write and then write some more. You should eat well and exercise. Ideally, you must also try to maintain a social life: go to the movies, see your friends. Read a book or two. Fiction, even. Some days, it feels like you've just enrolled in clown school, and on the first day the instructors have you trying to juggle three balls, a ferret, two knives, and a bowling pin. Then—don't duck!—here comes a chainsaw. (And, by the way, it's on fire. Don't kill the ferret.)

So why would you want to even entertain the notion of doing more work right now? Why on earth would you start your own blog while in grad school? Indeed, graduate students in the STEM fields (and all scientists, more generally) overwhelmingly blame time constraints for their inability to devote themselves to science communication.[1] The assumption—whether that's your assumption as a graduate student, your adviser's assumption, or your assumption regarding your adviser's assumptions—is that spending the time doing outreach will interfere with the goal of learning how to conduct research.

But that assumption, as you will see, is flawed. It turns out that if you plan your science communication efforts thoughtfully, then you can capitalize on them to build your own scientific skills and to advance your own scientific career. And that's not to mention the benefit of contributing to scientific literacy more generally among your friends, family, and the general public.

If you're not in a scientific program and are instead a graduate student in science journalism, you may think that you're already doing enough writing as it is. But although blogging may not sharpen your writing skills the way it might for a scientist, the benefits of blogging reach far beyond writing practice. Just because you're already writing a lot as a part of your course requirements and your internships doesn't mean you can't benefit from blogging as well. So while most of the remainder of this chapter approaches the question of blogging from the perspective of a student in a scientific program, keep in mind that many of these arguments also apply to you.

Blogging as Professional Development

One of my motivations for starting a blog in my second year as a graduate student in a psychology department was to practice writing about science. If I wrote a post or two a week, I thought, then when it came time to write my dissertation, the writing process itself would be far less daunting. Indeed, by the time I was approaching the end of my time in graduate school, I had been a science blogger for four years. I had written, give or take, some 1,500 to 3,000 words about science nearly every week. With all that practice, hammering out a dissertation's worth of words was a relative breeze. I was used to sitting down in front of a blank screen, typing and typing and typing until there were no more words left to type, then waking up the next day and doing the same thing. I wrote a first draft of my dissertation over the course of about three weeks, and spent another week or two editing it and making it pretty. When I started graduate school, putting together a doctoral thesis seemed like climbing Mount Everest: I'd need a team of sherpas, and there was a very real chance I might wind up dead. By the end, it was just another writing project, more like climbing the stair machine at the gym: a workout, to be sure, but something I could confidently complete without fearing for my very survival.

Here are three ways in which practicing science communication as a graduate student will help build the skills necessary for success in science or science communication.

WRITING

As my own experience illustrates, the best way to practice writing is to write, and with frequency. But blogging also widens your audience, which will enrich your writing even more. "As a matter of course," wrote Lauren M. Kuehne and her colleagues in a 2014 paper, "students may be expected to write articles for publication in scientific journals and give presentations at conferences, but the reach of those forms of communication is generally limited to other scientists in the field. By contrast, less traditional tools such as maintaining a personal website or contributing to a blog can improve writing skills while allowing non-scientists to access student research directly."

With a blog, you can learn how to build a compelling argument and, thanks to the comments section, receive nearly instant feedback. It doesn't matter if your commenters don't agree with you, as long as they understand your arguments in the first place. Blogging will teach you how to craft better analogies and stronger metaphors, simply because much of the best science writing is built around those techniques. Most people can't intuitively visualize particles in an atom, organelles in a cell, or neurons in a brain. As you write more you will learn how to describe the mysteries of the universe in concrete, easy-to-grasp ways. And by doing so, you will learn the importance of brevity. You will learn not to write in three paragraphs what could take only one sentence to clarify. When you're writing for your colleagues, you might take time to explain the complex chemosensory properties of the suckers that line octopus arms. But for a general audience, you might simply ask them to "imagine what it would be like if the majority of your body were made of tongues, able to both touch and taste the entire world," as I once did.[2]

Becoming a skilled science writer is more than writing good articles and blog posts. It's learning how to write clever, snappy headlines that will make readers more likely to click a link or to linger over a page. It's also figuring out how to best pair your words with photos and graphics.

As a scientist, the ability to communicate complex, nuanced ideas in more accessible ways is important beyond your ability to do outreach. That's because scientists in other fields aren't familiar with the jargon of your field, let alone the fundamental theories and assumptions that underlie your research—yet in many cases they will be responsible for evaluating your grant proposals, interviewing you for faculty jobs, and voting on your tenure application. Even if you don't care one bit about increasing scientific literacy among nonscientists, you should care about developing your ability to explain science to folks outside of your own field. Besides, if you can't explain your science in a way that a motivated high school student can understand, you might not really understand it that well yourself.

TEACHING

All those skills you develop by writing and writing and writing some more? Guess what—you need many of them for teaching too, which is also an important ability to strengthen, especially if you plan on remaining in the academic world.

Being a great science teacher is not so different from being a great science writer. You have to convince your audience to pay attention to you, rather than to the myriad other potential sources of entertainment and engagement out there. You then have to maintain their attention: at any time, a reader can click over to a different website or turn the page of the magazine. In a classroom, your students are just three seconds away from surfing over to Facebook on their laptops.

As a teacher or a writer, you also have to break down complex ideas into understandable chunks. Most importantly, the contract between teacher and student is a lot like the unspoken one between writer and reader. The better you learn how to write compelling prose, the better you will be at entertaining your students who, in many cases, might not actually want to be in class in the first place. The writer implicitly says to the reader: I will value you and your time, because I know that you are just two boring sentences away from reaching for your cell phone and playing *Candy Crush*, and in return you will spend some unspecified amount of time reading my sentences, equal to only a small fraction of the time that I spent crafting those sentences in the first place, and I will be thrilled to receive your attention, no matter how brief.

SPEAKING

Public speaking is a common fear. Improving your writing skills will no doubt help you give better, smoother, more professional talks. In addition to the teaching-related skills mentioned earlier, blogging will help you learn when to lean less on words and more on imagery. That's perhaps the most important thing to learn as you build your slide decks. Next time you attend a talk—perhaps a department lecture, or a job talk, or even a public talk—pay attention to the mechanics of the talk. Are the slides too text heavy? Is the content of the talk appropriate for the audience? Is the speaker allowing the audience time to examine the slides? Is the speech itself consistent with the content on the slides, or is the speaker saying one thing while the audience is distracted by looking at something else? Is there too much jargon? You will quickly see that many of the basic rules for creating effective blogs translate to public speaking.

As a scientist, you will be expected to give many talks: to your colleagues, to potential employers, and to the public. If you receive funding from a nonprofit group, giving talks may be part of the deal. One nonprofit in Los Angeles, for example, funds academic research on urban wildlife. As part of their agreement, scientists who receive funding are expected to give talks to update the members of the organization on their findings. And of course the ability to give talks without becoming anxious can be very useful when looking for a job. If your career path is professional science writing, public speaking may be less of an expectation, but writers are increasingly using speaking engagements to supplement their income. If you end up going on book tours and giving interviews on TV and on the radio, too, strong public-speaking skills will certainly be needed.

THE REST

There are plenty of other, smaller benefits that come from blogging about science while you're still in graduate school. When I talk to graduate students who are interested in blogging about science, the overwhelming stumbling block is that they feel like they don't have enough time. The truth is that all that time you spend during your workday checking Facebook and daydreaming about what sort of microwaved food product you'll eat that night (buy some vegetables and make a salad, you poor graduate student!) can be better utilized. Learning time management will set you up for success in your professional life, no matter what route you take.

Learning to blog will, almost by accident, mean that you will learn the basics of setting up a website. These days, all you need to set up a little corner of your own on the Internet are some very basic HTML skills and an understanding of how to operate a content management system like WordPress. Whatever career you wind up in, whether in science, journalism, or something else entirely, being able to manage your own Internet presence is an increasingly vital skill, and one that will make you even more attractive to potential employers.

Finally, if you're a scientist, you are no doubt aware that post-publication peer review is increasingly important. Conventional peer review is still an important mechanism for ensuring that the science that gets published in academic journals is coherent and accurate, but it is not perfect. No longer will journalists from traditional media outlets be the only ones writing about your work. Instead, bloggers—some of them scientists themselves—will be writing about your research on their own blogs. Even if you don't intend to become a prolific blogger yourself, understanding more about the culture of science blogging will allow you to respond appropriately when your work is either praised or criticized. Having your own blog will also, of course, give you a platform for communicating your own science yourself. That's especially important if you feel that the mainstream media have overstepped in their interpretation of your findings, or if you feel that your colleagues have unfairly criticized your work. Like any other community, online science communication has its own culture, its own set of spoken and unspoken rules. The more familiar you are with them, the better you can engage productively with your colleagues.

Finally and most importantly, you should blog because it's fun. What other excuse do you have to read obscure papers in your field that aren't really relevant to your own research and then tell the world about them? In what other venue can you lean on animated cat GIFs to explain some complicated concept in your field? Let's face it: if you're reading this book, you're probably thinking about blogging already. And if you're in grad school, you need all the entertainment you can get.

JASON G. GOLDMAN is a freelance science writer. He has written for *Scientific American, Los Angeles Magazine,* the *Washington Post,* the *Guardian, Slate, Salon, io9, BBC Earth,* and elsewhere. He contributes multiple pieces each week to *Conservation Magazine,* is a columnist for *BBC Future,* and is a staff writer for the *Earth Touch*

News Network. He also hosts a weekly podcast called *The Wild Life,* sponsored by *Earth Touch.*

Jason is based in Los Angeles. Find him at his website, http://www.jasonggoldman.com, or follow him on Twitter, @jgold85.

Notes

1. Lauren M. Kuehne et al., "Practical Science Communication Strategies for Graduate Students," *Conservation Biology* 28, no. 5 (2014): 1225–1235.

2. Jason G. Goldman, "The Alien Brains Living on Earth," *Uniquely Human, BBC Future,* June 26, 2014, http://www.bbc.com/future/story/20140626-the-alien-brains-living-on-earth.

18

Blogging on the Tenure Track

GREG GBUR

Faculty striving for tenure often experience the greatest pressures of their careers: pressure to achieve in the laboratory and the classroom, as well as pressure to do outreach. How can writing your own science blog help you? Could it be the icing on the cake of your tenure package? Greg Gbur explores the advantages and difficulties of blogging while on the tenure track.

Scientists pursuing the tenure track in academia have a lot to worry about: they must supervise students, conduct research, write papers, teach classes, help manage the department, and get funding for their research, among many other commitments. It may seem incredible, but plenty of academics choose to take up blogging on top of all these tasks. Others have thought about starting a blog, but wonder if there are any benefits to it (besides the satisfaction of reaching out to the public) that can outweigh the risks to tenure. With that in mind, I aim to explain why one might want to blog on the tenure track and, equally important, to suggest how to do it while minimizing risks to one's tenure case.

I started my blog *Skulls in the Stars* back in 2007, roughly two years into my position as an assistant professor of physics in North Carolina. In the beginning, I ran the blog pseudonymously, though I never worked too hard to hide my activity. To receive tenure, one must typically put together a package of documents showing one's strong contributions to research and teaching, as well as service to the university and one's professional community. When I went up for tenure myself in 2009, I included a description of my blogging activities in both the "teaching" and "service" portions of my tenure package. The complete document was well received, and even used as an example for other assistant professors in the following year. Since then, blogging and social media have become an accepted and rather important part of my academic life. Some of the advice I'll share stems from plans I put in place before tenure, and other ideas occurred only in hindsight, or were accidental!

How

Let's start with the "how" discussion first. Suppose a new faculty member decides to write a blog, or has an already established blog before taking the new job; how should he or she make sure it is treated as a benefit and not a detriment to the university? This is not an unreasonable concern: though blogging and other social media activities have become much more accepted even in the last five years, a new professor still runs the risk of having it viewed as an activity that takes away from the already heavy load of important work that needs doing. One's overall strategy should involve showing officials at all levels of the university that blogging adds to academic life, and doesn't detract from it.

To achieve this goal, make sure that your official duties take priority over blogging, and that your "research," "teaching," and "service" components of your tenure package are beyond reproach. This is somewhat obvious, but easy to forget in the enthusiasm of writing fun blog posts. I tend to view blogging as a particularly delectable dessert that comes after a tenure package dinner: the dessert can't take the place of the meal, but can add to a particularly good one.

Closely related to this is making it clear that blogging is an extracurricular activity, and not one taking away from ordinary work time. I emphasized this in both word and deed. On the "deed" side, my online activities were mainly done at night, after work, for the first couple of years. Since getting tenure and recognition for my efforts from the university, I've felt more comfortable developing blog posts during work hours. On the "word" side, I would occasionally mention in blog posts and department activities reports that my blogging was an after-hours effort.

Tenure decisions are made by one's colleagues, beginning with a decision by an appointed RPT (reappointment, promotion, and tenure) committee and often including a vote by the entire department. This means that it is highly important to connect with departmental colleagues and make sure they understand whatever you're doing, including blogging. In the (thankfully few) cases where I've seen faculty fail to get reappointed or to get tenure, they've typically been very isolated from the rest of the department. Such faculty members are unable to find allies because either nobody has a vested interest in their work, or nobody is even sure what they've been doing. I told my colleagues about my blogging gradually, starting with those I was friendliest with and working my way through the rest of the department. I also invited a number of colleagues to guest blog for me, which was not especially successful, but making the effort to be inclusive may have made people more sympathetic to the idea of blogging in general.

Department colleagues are far from the only allies to be found on campus. All universities have public relations offices that are tasked with promoting the university, its research, and its faculty. Traditionally, communications workers in these offices have served as intermediaries between the university faculty and the public by, for instance, writing press releases about exciting research developments. Getting faculty to help with this process, however, is often like pulling teeth, and I've found that workers in communications are delighted to find scientists who are active promoters. Helping with a university's social media strategy is a great way to make oneself a unique and indispensable asset.

There are many different ways a faculty member can help with outreach beyond promoting one's own research. When word of my blogging got around, I was approached by a research communications specialist about organizing and running a social media workshop sponsored by the College of Liberal Arts and Science. This was done pre-tenure, and gave me some official sanction for my blogging efforts, which in turn made me more confident about putting blogging in my tenure package. In recent years, I've also volunteered to help with my university's yearly Science and Technology Expo, a free event open to the community (http://cri.uncc.edu/research-innovation/2014-science-technology-expo). It can be worthwhile, too, to see if one's university has a series of popular talks for the general public, and volunteer to give one. At the very least, I would recommend contacting the university's public relations office and finding out how your work can fit into the university's larger promotional strategy.

It is worthwhile to connect with people outside one's university as well. It is now possible to find established researchers blogging about almost every conceivable field of study, and I suggest getting to know people in one's field by interacting through blog comments, Twitter, and other social networks. Social media is an excellent place to ask questions and share ideas, and good conversations online can leave a positive and lasting impression on colleagues. Not only can these researchers act as virtual mentors for one's own online activities; they can also be sympathetic referees on one's final tenure package. For those people, like me, who find meeting people at professional conferences awkward, an initial introduction online can smooth the process considerably. Just be sure to be polite, sincere, and not too pushy in your online interactions, especially with very distinguished researchers who may already be stretched thin with a large number of other commitments.

It is also important to connect with professional organizations. Many organizations have now jumped wholeheartedly onto the social media bandwagon, and are very eager to find members who are themselves active online. Connections with these groups can add professional gravitas to one's online activities; in my final tenure package, I was able to note that my blog was listed on the blogroll of the American Physical Society's own blog, *Physics Buzz*. As a bonus, professional organizations will occasionally offer registration discounts or other incentives for participants willing to be official bloggers for a meeting.

Connecting with events, organizations, and activities specifically attuned to online science communication will help give one's efforts an official appearance. There are now numerous physical meetings about online science outreach on the local, regional, and global levels; participating in these meetings or even presiding over a session is a nicely tangible accomplishment.[1] A number of professional organizations and science magazines offer opportunities to write guest blog posts, which can bring in more readership to one's own blog as well as work as a semi-official publication. Though they seem to be gradually going out of style, some "blog carnivals" are still going that one can volunteer to host on one's site; these carnivals are monthly or semi-monthly collections of links to posts on a particular topic. Participating in a carnival will demonstrate that a faculty member is part of a broader community of writers. I myself, with some encouragement, started a blog carnival about the history of science known as *The Giant's Shoulders* (http://ontheshouldersofgiants.wordpress.com), which ran for six years and which I counted as an achievement in my tenure package.

It should go without saying that all achievements of this sort should be documented for tenure. Blog statistics, especially numbers of blog comments, are important metrics that demonstrate the impact of one's efforts. Even modest stats, say two hundred page views a week for a year, add up to numbers that are quite impressive in comparison with ordinary academic citations. Very few researchers can claim to have had their papers read more than ten thousand times before tenure, though a single blog post can reach that number very quickly!

Selling blogging to an academic department becomes easier as one gets more established online and one's influence becomes greater. For those who are more cautious, it is not unreasonable to begin blogging under a pseudonym, which provides some level of protection from online gaffes and allows the faculty member to step away easily if the activity doesn't feel right. I started my own blog under a pseudonym, though by the second year I felt established enough to share my work with colleagues (I realized, too, that one must inevitably come out from hiding if one wants credit.) Keep in mind, though, that no pseudonym offers perfect protection from being "outed": one should always keep blog behavior in accordance with acceptable university practice. The best guideline: if you wouldn't say it in a professional setting, be wary about saying it online.[2]

Why

Why would a faculty member want to put in the effort, on top of his or her already busy workload, of creating and maintaining a blog? Hopefully the most important answer is that the researcher genuinely enjoys connecting with people online about science. Beyond this, however, there are a number of other practical benefits to blogging on the tenure track.

I began blogging with one significant goal in mind: broadening my scientific knowledge. Having a blog is great motivation for reading about research in not only one's own field but also areas outside of one's immediate area of expertise. Reading about a wider variety of research topics can in turn give a scientist new ideas and lead to novel research. I am now developing a pair of small research projects that were motivated by reading for my blog.

Writing publicly about other people's research is also a great way to make new scientific connections and, potentially, find new collaborators. Scientists are typically delighted to read positive and accurate descriptions of their work, and it is not unusual for them to comment on the blog or contact a writer directly about such a post. (Keep this in mind if you decide to write something harshly critical.) Demonstrating a deep understanding and enthusiasm for a scientist's research also makes the possibility of collaboration more likely; at the very least, it puts a blogger "on the radar" of the scientist. I personally have not yet had any significant blog-based collaborations, but a number of esteemed scientists are now aware of my research after first encountering my blog writing. Such familiarity can lead to opportunities for presenting research at conferences, for helping to organize conferences, and for journal editing, among other things.

Perhaps most important, blogging has been a great platform for improving the quality of my own writing, and it comes with free and honest feedback in the form of comments on blogs and other social media. Seven years of blogging have taught me how to write and explain complicated science in a nontechnical manner: my early posts are on simple topics and filled with equations, while my more recent posts are on complicated topics and contain no equations. Being able to talk science in an entertaining and nontechnical way is of course a very useful skill in the classroom; what surprised me is how much better my technical papers and conference presentations became as well. Successful talks and papers should always have an introduction understandable to the broadest audience possible; my blogging work has helped me write more accessibly.

Improving your writing skills may produce its own tangible benefits: it can lead to offers to write popular science articles for magazines and possibly even books. Thanks to my blogging efforts, I have written several articles for *Optics and Photonics News* (http://www.osa-opn.org/home) and the French magazine *La Recherche* (http://www.larecherche.fr). Both opportunities came from editors who read my blog. There is a real need, especially in professional science organizations, for scientists who can write well about technical topics, and blogging is a way to freely advertise one's writing skill. Plenty of writers have also used their blog as a platform to develop, advertise, and sell a book proposal: because of posts I have written, a publisher has expressed interest in working together on a book.

Blogging, and online communication in general, have also become remarkable tools for sharing and solving problems as a community. On Twitter, where "followings" can connect you to thousands of people at once, it is possible to get answers to questions, assistance finding research papers inaccessible at a particular institution, or even help solving a scientific mystery. To take just one example of many: Alex Wild, who blogs at *Myrmecos* (http://myrmecos.net/2012/02/01/an-unusual-wasp), took a picture in early 2012 of an unusual-looking wasp in Australia. Unable to identify it, he posted the picture on Twitter, and within twenty-four hours learned via the North Carolina State University Insect Museum that he had taken what may be the only live photographs of that insect.

A few other surprising advantages of blogging are worth mentioning. First, it can push academic scientists into entirely new and unexpected fields of study. When I started blogging, I had a mild interest in the history of science, but the response and feedback from early posts led me to become actively involved in its study, and I am now accepted as an amateur historian of science among the professionals. Second, and perhaps most delightfully, blogging allows academics to connect in a less formal manner with students and the public. The first panel discussion on blogging that I co-moderated, for example, was with an undergraduate and a graduate student. We were equals in running the session—and I learned a lot. Having an online presence helps a scientist keep one foot outside of the "Ivory Tower," making him, or her, a better public academic. In an age when science is often attacked as being unimportant or untrue, it is important for academic scientists to be able to explain and justify their own work to people of all backgrounds and professions.

If carefully and thoughtfully orchestrated, blogging can give a faculty member on the tenure track the perfect training for interacting with people at all levels inside and outside of academia. A dedicated blogger becomes an outreach resource for the university—a faculty member who can communicate in a way that most cannot. While blogging on the tenure track, your goal should be to convince faculty and administration at all levels that you can fulfill that role.

GREG GBUR is an associate professor of physics and optical science at the University of North Carolina at Charlotte. He writes about physics, optics, and the history of science at his blog *Skulls in the Stars*. His blog writing has appeared in *The Open Laboratory* in 2010 and 2013 as well as in *The Best Science Writing Online* in 2012.

Greg is based in Charlotte, N.C. Find him at his website, http://skullsinthestars.com, or follow him on Twitter, @drskyskull.

Notes

1. See, for instance, the long-running ScienceOnline series of meetings, http://scienceonline.com, and some of its offshoots such as ScienceOnline Oceans and ScienceOnline Climate.

2. The cases of David Guth (http://kansasfirstnews.com/2014/04/02/ku-professor-returning-after-leave-over-controversal-tweet) and Steven Salaita (http://kansasfirstnews.com/2014/04/02/ku-professor-returning-after-leave-over-controversal tweet) are noteworthy: both have had their careers suffer due to angry statements on Twitter.

19

Metrics

Measuring the Success of Your Blog

MATT SHIPMAN

Most people start communicating about science on the Internet because they want to reach a wide audience. But now that comment threads have moved to Twitter and Facebook, it's becoming harder to tell who is seeing your work and what kind of influence you have. Matt Shipman is a public information officer at North Carolina State University and blogger at Nature Publishing Group's scilogs.com. *He will explain how to tell if there's anyone out there, and if they are listening.*

So you started a blog to give yourself a place to communicate with people about science. And hopefully you started the blog with specific goals in mind. You may want to inspire a new generation of astronauts, raise awareness about the importance of the microbial life that surrounds us, or simply share an online diary of your field expeditions in the frozen Arctic or the Arizona desert. Whatever your intentions, you need to have some way of determining whether you are meeting those goals, whatever they are. In short, you need metrics.

Metrics are simply tools that you can use to measure, well, anything. But you will want to focus on those metrics that actually measure progress toward your goals. It's easy to get overwhelmed by numbers. There are a lot of different things you can track—from how long people spend reading your work to how many people are sharing it—and it's important to figure out which numbers are important to you and how you can use that information to share your work more effectively.

In this chapter I will explain the nuts and bolts of collecting data, what those data can tell you, and how you can use that information. I will also highlight the importance of creativity and critical thinking when it comes to measuring success. It may be easy to track numbers, but numbers alone can be misleading. You probably didn't start writing or making videos with the ultimate goal of getting a thousand viewers or of having people spend more than ninety seconds on your landing page.

Most people get into science communication because they love science. They want to change the way people think about scientists, educate them about scientific subjects, encourage them to become scientists, or get them involved in citizen science projects. Numbers alone won't tell you if you're achieving those sorts of goals. So while it's good to keep track of the numbers, it's important to come up with additional metrics that will help you determine whether you are accomplishing what you set out to do in the first place.

Conventional Metrics

If you're managing a blog, you probably have some fundamental questions: How many people are visiting my blog? Which posts are most popular? Where did these people find my blog in the first place? The best way to collect this sort of information is by using analytic tools that are provided by your blogging service or that you can add to the blog yourself.

Analytic tools give you a variety of metrics about how many people are visiting your blog, where they came from online, and which of your pages those visitors are viewing. While the terminology may differ slightly from program to program, I'm going to give a basic overview of a few of these metrics.

When looking at analytics for your overall blog, most programs will tell you how many "visitors" and "unique visitors" you have. "Visitors" refers to the total number of times anyone has visited your blog, while "unique visitors" refers to how many specific individuals have visited your blog. For example, if one person visits your blog every day for a week, that will show up as seven visitors but only one unique visitor.

Similarly, most analytic programs will also give you information on "page views" and "unique page views." "Page views" refers to the number of times that anyone has visited a specific page, or post, on your blog, while "unique page views" refers to how many specific individuals have visited that page. So if one person opens your website and clicks the refresh button a thousand times, you'll have a thousand page views, but only one unique page view.

These metrics are useful for tracking overall traffic to your site, and you can use them to make editorial decisions regarding what you write about and how to write about it (or, for multimedia folks, what sorts of art or videos resonate with your audience). For example, if the metrics tell you that 50 percent of all the visits to your site were for a post you wrote about bioluminescence, you might want to write about bioluminescence again in the future. By the same token, if posts written in conversational language draw more visitors than posts that incorporate technical jargon, you may want to increase your efforts to write in casual language.

Two other metrics that can influence your editorial decision-making are "bounce rate" and "average time on page." Bounce rate tells you what percentage of visitors clicked a link to go directly to a specific page on your site and then exited the site without viewing any of your other pages. Knowing this can be incredibly useful. For example, if there's a page that stands out as having a low bounce rate, it's worth taking a closer look. What distinguishes that page from your other posts? Did you include hyperlinks to other pages in the body of the post? Did you do a good job of highlighting related posts in a sidebar or "related stories" box? Figure out what you did that kept people on your site, and try to incorporate those techniques into future posts.

"Average time on page" tells you how much time, on average, your visitors are spending on a specific post. If people are spending minutes on the page, that means they're probably reading it. But if people are spending only fifteen seconds on the page, you should investigate, because it means that something is scaring them off. As with bounce rate, see if you can figure out what is different about that page and adjust your overall strategy accordingly. For example, if posts that use technical jargon in the first sentence are scaring off readers in less than twenty seconds, it's a clue that you need to refine your writing technique. After all, you (presumably) want people to read what you wrote—and you can always introduce the technical language further down in the post.

Analytic programs can also help you determine *how* you are reaching readers. This information goes by different names in different analytic programs, from "source" to "transitions" data. But all of it tells you how people arrived at a specific page on your blog. For example, it might indicate that 10 percent of the people coming to the page came from a Reddit post, 15 percent from Facebook, and so on. This sort of "traffic source" data is useful in two ways. First, it can tell you about activity related to your page that you didn't know about. For example, I usually don't find out that reporters have linked to my blog posts until I see it in my traffic source data. Seeing the stories that are related to my blog post can help me to evaluate the subject from a different angle—and so ultimately inform or inspire my future blog posts.

Second, and more importantly, traffic source data let me know which avenues are most effective for disseminating my blog posts. For example, if I find that the bulk of my readers are coming to my blog from Twitter, that tells me I need to maintain my presence on Twitter. Conversely, if I'm spending a lot of time and effort trying to push out my posts on Twitter, but getting very little traffic, I need either to change my approach or to focus my time and energy elsewhere.

Social Media Metrics

To determine whether you are making progress toward meeting your goals for the blog, you need to know whether you're reaching your target audience. And as useful as blog-specific analytic tools are, they don't tell you who your readers are. For clues to that, turn to social media.

I think Twitter is the most effective social media platform for disseminating blog posts. In part, this is because Twitter is home to a wide variety of discipline-specific networks in which people interested in a specific field share links to relevant material. This is particularly true within the sciences. While these networks are usually informal, they are very effective ways of reaching a lot of people with shared interests.

To begin tracking whom you reach on Twitter, simply click on the "notifications" tab. This will show you every tweet that mentions your Twitter handle, as well as any "retweets" (or shares) and "favorites." But the notifications tab won't tell you anything about tweets that link to your blog if those tweets don't include your Twitter handle. Luckily there are other tools to help you do that. I'm particularly fond of services like Topsy.com that show you everyone who has shared a link to a given page. If you can see who is disseminating your work, you can get a good idea of whether you're reaching your target audience.

Facebook is more problematic. It is undeniably useful for sharing information with friends and family. If you create a page for your blog, however, it is no longer reliably useful as a tool for sharing information with your followers, thanks to the vagaries of Facebook's news feed algorithms. In particular, most Facebook pages are able to reach only a tiny fraction of their followers with any given post. Unless you have the budget to "promote" your Facebook posts, then, I recommend that you focus your social media efforts elsewhere.

Unconventional Metrics

There are a lot of social media platforms out there, and dozens (if not hundreds) of social media metrics tools. In addition, both the platforms and the metrics tools are changing all the time. You have to stay current and be willing to experiment with these tools to find what works best for you. It also pays to remember that although the universe of metrics tools is evolving quickly, there are two tools that will always work: creativity and critical thinking.

Let me explain. As I mentioned earlier, to have a successful blog you first need to have a clear idea of what you're trying to accomplish. Once you've set clear goals, you can bring critical thinking to bear and figure out what you need to measure to determine whether you're making progress toward those goals. And once you know what you need to measure, you can use critical thinking (again) to figure out creative ways to measure it.

The conventional metrics I described earlier—analytics and social media—are useful. But developing your own unconventional metrics may be even more useful for determining whether you've made progress toward your specific goals.

I'll give you an example. I work as a science writer at a university, and I often write for the university's research blog. The primary goal of the blog is to raise the profile of the university in a positive way by highlighting the work done by faculty and students. A great way of doing that is to write stories for our blog that we think can be picked up by news media. We want people to visit our blog and read what we've written, but we know that stories that run in external news outlets will reach a much broader audience. As a result, while we do track visitors to our blog, an unconventional metric that we use to measure success is the number of external news stories that a blog post generates. We also track social media mentions of our posts, because they can serve as indicators of interest in the work we write about (because, presumably, people won't share links to a post they don't find interesting).

Another objective of the university's research blog is to help our faculty and students achieve their own goals. Secondary goals like these often require their own unconventional metrics. A case in point would be a post I wrote in 2013 about a research assistant at the university who had recently earned his undergraduate degree.[1] The assistant had begun working on a product in his senior year and, after graduating, had teamed up with another inventor to bring the product to market. He and his partner had set up a Kickstarter page to fund an initial production run, and I was contacted about the work he was doing.

I wanted to write about the research assistant because it would be a good story for the university, highlighting student ingenuity and entrepreneurship. The research assistant wanted us to write about it because he wanted to draw attention to his fledgling company and generate interest in his company's Kickstarter campaign.

I wrote the story, posted it on social media sites, and brought it to the attention of a few journalists whom I thought might be interested. And then I used two sets of metrics to determine whether the post was a "success."

The post led to more than a dozen stories in external news outlets, including the largest newspaper in our metropolitan area. It was also shared more than a thousand times on social media. That means the university successfully highlighted student entrepreneurship to a large audience, and that the audience was extremely interested. In other words, the post accomplished the primary goal we had set for the blog.

Did we also achieve our secondary goal? Yes. And one of the unconventional metrics we used to determine whether we were successful was the amount of traffic we drove to the research assistant's site. The assistant reported that there had been a huge jump in visitors, which was good—we had stirred up interest in his project. The second unconventional metric was activity on his project's Kickstarter page. Donations jumped by more than $200,000 after our blog post (and the ensuing media coverage).

In short, the post did exactly what we hoped it would do: it raised the university's profile in a positive way and helped a fellow employee or student achieve his or her goals. We had the unconventional metrics to prove it. Significantly, if we had relied only on conventional metrics, the post would have looked like a failure. The post received only 3,800 unique visitors and had a bounce rate of more than 93 percent.

The moral of the story is that conventional metrics are great, but don't rely on them to tell you whether you're reaching the goals you've set for yourself. It is worth taking the time to come up with metrics that give you specific, meaningful insights into whether you're doing what you set out to do.

MATT SHIPMAN is a science writer and public information officer at North Carolina State University. He writes the *Communication Breakdown* blog for *SciLogs,* and is the author of *The Handbook for Science Public Information Officers* (University of Chicago Press, 2015).

Matt is based in Raleigh, N.C. Find him at his blog, http://www.scilogs.com/communication_breakdown/author/shipman, or follow him on Twitter, @shiplives.

Note

1. Matt Shipman, "Science You Can Use: Engineer Designs Mug to Keep Coffee Temperature Just Right," *The Abstract* (blog), North Carolina State University, December 11, 2013, https://news.ncsu.edu/2013/12/wms-hot-coffee.

20

Toot Your Own Horn

Self-Promotion in Social Media

LIZ NEELEY

Many writers have been raised to believe that their work should stand on its own, that self promotion is both self-centered and counterproductive. But on the Internet, these activities are not vanity but a necessity. Liz Neeley, executive director of The Story Collider, *will show you the power of self-promotion.*

Congratulations! That is a solid post you just finished. It may not be perfect, but it is done, and you should feel good. Nobody writes a research blog about topics they don't care about, so let's not make light of the hours (days? weeks?) of work you have invested to this point. But publishing your post is just the beginning. Now it's time to promote it.

Ugh, self-promotion. Everyone knows those people, the relentless self-promoters, and everyone knows how we tend to describe them. But what they do works. You need to get your ideas in front of as many interested people as possible, and more importantly, you want to get your ideas in front of us in such a way that we will read, remember, and respond to them. You need our attention, but our attention is an incredibly valuable resource.[1] Consider the multitude of delicious (baking cookies), important (writing), or helpful (walking my dogs) things I could be doing at this moment. Reading your post comes with a price. If I choose to read it, I have chosen not to do all those alternatives. If I open up your post and start to read it, I've already made a calculation in your favor—I believe that what I'm about to read is going to be worth the opportunity cost. Maybe it's because whatever I saw (a tweet, a Facebook post, perhaps simply the post title) piqued my interest. Maybe I'm basing my decision on my knowledge of other things you've written. Either way, your self-promotion is a promise to me, one of your potential readers, that your work will be worth my attention.[2]

Why Self-Promote?

As with any communication challenge, before you can effectively decide about the "how" of self-promotion, you have to design a strategy that clarifies the "why." There is a wide range of personal motivations for getting into science blogging, but broadly speaking, the ultimate goals usually include influencing opinions, driving the social agenda, and advancing our collective knowledge.[3] Whether you are blogging about sophisticated theories of change and logic models, or keeping things relatively simple, you'll need a clear understanding of why you are blogging, who will likely read each post, and what you want your readers to do afterward.[4] Strategic self-promotion is critical for achieving these proximate goals in your overall strategy: members of your target audience cannot read your work, for example, if they don't know you exist. It can also bring indirect benefits like positioning you as an authority and raising your professional profile, both of which can trigger a cascade of more and better invitations to speak, collaborate, and advance your career.

Self-promotion is necessary. But for many of us, there is a deep-seated belief that while the work of content creation is noble, the work of drawing attention to your content is distasteful if not degrading. Our aversion to self-promotion is an emotional reaction, exacerbated by the suspicion (particularly if you are female) that the usual advice for increasing traffic—repetition, jumping into comment threads to mention your post, direct requests to retweet—can annoy the very people you hope to impress.[5] In the past five years, I've taught social media to hundreds of researchers in dozens of workshops, and I have never had a discussion about self-promotion that didn't feel at least a little uncomfortable. To reassure researchers about how reasonable and utterly normal this is, I've made a series of slides with my Google search results when I start typing the phrase, "Promote yourself . . ." The first five suggested autocompletions include "without being sleazy," "without being a jerk," and "without talking about yourself."[6] No one wants to look like a selfish, self-involved jerk, but early successes tend to snowball, so the cost of keeping quiet compounds over time. As my mom said to me on the eve of my high-stakes election for sixth-grade class government, "Honey, if you don't vote for yourself, why should anybody else?" The bottom line: when it comes to self-promotion online, refusing to engage is not taking the moral high ground, it is self-sabotage.

The "How"

Right now, the biggest players in the science communication landscape are blogs, Facebook, Twitter, Google+, Tumblr, Reddit, and YouTube, and they all provide fertile ground for research. The emerging field of content virality is fascinating.[7] It's interesting (and maybe even important) to understand why Upworthy headlines are so successful, or what time of day is best for maximizing retweets.[8] Yet it seems that we are best at discussing what precise combinations of Internet architecture, social network topologies, and the human brain have combined to make something blow up only in hindsight. We don't know how to deliberately design viral success, despite what the gurus and mavens may say.[9] Even worse, no one can guarantee which platforms and networks are going to offer the best return on your investment. Success seems to be a singular history of trial, error, luck, personal preference, style, and timing.

To take just a few examples, while I was writing this chapter, there was a portentous shift of staff and resources at Google+, and a high-profile piece in the *Atlantic* eulogizing Twitter.[10] Within the month of this writing, Twitter redesigned its user profiles, while Facebook and YouTube have updated their looks and algorithms in ways big and small.[11] We are in the midst of a period of rapid change, and what it means for self-promotion is hard to say. If we can stay flexible and hungry, we may find it easier to navigate each new wave and to seize first-mover advantages. But maybe we just have to accept that we're in the midst of a great uncertainty. I'm reminded of this quote from Clay Shirky: "[People] are demanding to be told that old systems won't break before new systems are in place. They are demanding to be told that ancient social bargains aren't in peril, that core institutions will be spared, that new methods of spreading information will improve previous practice rather than upending it. They are demanding to be lied to."[12]

We must approach outreach efforts with the same skepticism and rigor that we bring to science in the first place. That means embracing an experimental mindset. I'm inspired by Portland (Oregon) art and design firm Wieden + Kennedy, which emphasizes both the essential role of taking creative risks (to generate new ideas) and rapid prototyping (to test those ideas and move on quickly if they aren't good). In its entryway, a glittering wall of pushpins traces out *FAIL HARDER* while a mannequin with a blender for a head exhorts "Walk in stupid every morning." In interviews, founder Dan Wieden explains, "While you were sleeping, the world you're now inhabiting has changed somehow. It might be a big change, a small change, but don't assume anything. . . . Find out what's going on with your partners, with clients."[13] In your promotions, as well as your work, use the best available data to stay fresh and relevant.

How? Start taking advantage of free Google Analytics to get a handle on where your traffic is coming from. People are going to come across your post by one of two distinct paths: intentional search or serendipitous discovery.[14] It's a discovery problem, whether your audience knows they are looking for your content or not, so in either instance, promotion across multiple social media networks will help you cast a wider net. You can do basic A versus B testing with minimal effort (for instance, by simply employing different styles of tweets), and there are plenty of great resources for building a dashboard.[15] At a minimum, Google Analytics on your blog can help you decide where you want to focus your efforts, and whether you're seeing any differences as a consequence. Hypothesize. Test. Adapt. Rinse and repeat as you experiment with the following advice.

BE AUTHENTIC

This book has a chapter about finding your voice: go (re)read it. Now apply that same thinking to your promotional efforts. Play to your strengths. If you're funny, be funny! If you're not . . . please don't try. It's exhausting to attempt to be something you're not. Authenticity is an important component of likability and credibility in online interactions.[16] Earlier I described self-promotion as making a promise to readers. At COMPASS one of our touchstones is the saying, "You only get in trouble with people when you f*ck with their expectations." In short, your promotion should be a reflection of your authentic and unique style. There's no single proper way to do this, so you have to be true to yourself.

FOCUS ON CONVERSATIONS

A simple way to start pointing people to your work is to find those who are already searching for it or talking about similar topics. Search for keywords and jump into open discussions: if you're an ecologist, for example, find other ecologists, follow and comment on their blogs, and start conversations with them on Twitter. The basic premise of conversations is that they include both listening and contributing. Instead of shouting "Read my stuff! Read my stuff!" start by figuring out who might be happy to learn about it. Each platform has slightly different mechanics and norms, but at a minimum, you want to be using the right hashtags and following the correct circles/subreddits/pages/groups.

You don't have to wait for a question to be asked directly of you. You also don't have to confine yourself to your most recent post. Even in the relentless churn of content online, freshness is trumped by great material, and sometimes the valuable resource is not a post at all, but you. One of my favorite examples of natural and useful self-promotion is this tweet from Jacquelyn Gill: "@edyong209 @carlzimmer @Laelaps I've got an alternative hypothesis about this paper if you're writing it up! http://www.nature.com/nature/journal/v506/n7486/pdf/nature12921.pdf."[17] She knew that Ed, Carl, and Brian might cover a new *Nature* paper, and that journalists are always looking for expert critique and commentary on a new finding. Timely, topical, helpful: it's on point. Sometimes it works, sometimes it doesn't, but you'll never get better at it if you don't take the chances as they arise.

ADD VALUE

Think bigger than just your own content. You might find that you can better serve the topics you care about by starting a hashtag, tweeting other people's posts, or moderating a community. Similarly, you might Storify a great conversation, a Google+ hangout, or an event. Curation may be an exhausted buzzword, but it is a fundamental way to demonstrate credibility, structure complex information, and highlight material with lasting utility. Plus it is faster than content creation, and it tends to spark new ideas, so it can be a viable strategy for staying active when you're swamped and for staying fresh when you're less inspired.[18]

THINK ABOUT YOUR NETWORK

The self-promotion dilemma pops up when we don't want to be "that person" who is self-aggrandizing, but we know that likable modesty sacrifices attention and career opportunities. The science-based solution is to cultivate networks of people who will promote your work for and with you.[19] Having third parties speak and write on your behalf will increase your perceived likability and competence, while boosting your visibility in networks beyond your own.

You might activate your network by doing something as minimal as including the phrase "please retweet." There is some evidence that simple tweaks in phrasing can boost considerably retweets and click-throughs.[20] Similarly, asking influential users to share content can give you a huge boost. You should use this tactic sparingly, however. Is it an important post to you? Is the person someone you know well and/or someone with whom you share a deep interest in the topic? Then it makes sense to ask for support. Otherwise, perhaps it is better to ask for feedback, to link to or mention those doing related work, or otherwise to deepen the broader conversation. This blended approach is more effective than relentlessly pushing your own stuff.

For bloggers in particular, credibility is built as much through a kind of networked authority as it is through personal expertise.[21] In other words, no blogger is an island, so be a good citizen. Say thank you. Acknowledge those whose work you use or who have inspired you, and reciprocate when possible. Sure, page views and mentions are important, but these are really only proxies for something else. That something else is social capital: the goodwill, sympathy, and trust that manifest in support and action. This is the ultimate goal of self-promotion.

PROMOTE THE MESSAGE

Self-promotion is not a way of asking readers to do you a favor. We are all hungry for great content, so you are truly doing us a favor by helping us find yours. More than a year ago, I wrote, "'So Tweet This, Maybe?'—Promoting Your Work in Social Media."[22] I use the post in my teaching and I don't shy away from linking back to it when appropriate.[23] I don't know whether it is *the* reason I was invited to contribute this chapter, but I'm certain it helped. That's the power of self-promotion. Done well, self-promotion is acting in service of your ideas, not just clamoring for affirmation. Finding your voice, focusing on great content, and positioning it effectively can create positive spirals that benefit your work and your career. You have great ideas. Get over yourself, get out there, and help us discover them.

LIZ NEELEY is the executive director of *The Story Collider*. Prior to that, she was the assistant director of science outreach for COMPASS, creating and leading science communication training sessions around the country that emphasize both traditional and social media. She has also taught science communication at the University of Washington.

Liz is based in Seattle. Find her on her website at http://www.lizneeley.com, or follow her on Twitter, @LizNeeley.

Notes

1. Michael H. Goldhaber, "Attention Shoppers!," *Wired*, December 1997, http://archive.wired.com/wired/archive/5.12/es_attention.html

2. Glenn Llopis, "Personal Branding Is a Leadership Requirement, Not a Self-Promotional Campaign," *Forbes*, April 8, 2013, http://www.forbes.com/sites/glennllopis/2013/04/08/personal-branding-is-a-leadership-requirement-not-a-self-promotion-campaign.

3. Inger Mewburn and Pat Thomson, "Why Do Academics Blog? An Analysis of Audience, Purposes and Challenges," *Studies in Higher Education* 38, no. 8 (2014): 1105–1119, doi:10.1080/03075079.2013.835624; Hauke Riesch and Jonathan Mendel, "Science Blogging: Networks, Boundaries and Limitations," *Science as Culture* 23, no. 1 (2014): 51–72, doi:10.1080/09505431.2013.801420; Marie-Claire Shanahan, "Science Blogs as Boundary Layers: Creating and Understanding New Writer and Reader Interactions through Science Blogging," *Journalism* 12, no. 7 (2011): 903–919, doi: 10.1177/1464884911412844.

4. "What Is Theory of Change?," Center for Theory of Change, accessed February 13, 2015, https://www.theoryofchange.org/what-is-theory-of-change.

5. For three very different but complementary perspectives, try this combination: Ninja_OT, comment on Reddit, http://www.reddit.com/r/AskReddit/comments/1eysyr/whats_the_most_annoying_thing_about_social; Laura E. Buffardi and W. Keith Campbell, "Narcissism and Social Networking Web Sites," *Personality and Social Psychology Bulletin* 34, no. 10 (2008): 1303–1314, doi: 10.1177/0146167208320061; Janie M. Harden Fritz, "How Do I Dislike Thee? Let Me Count the Ways: Constructing Impressions of Troublesome Others at Work," *Management Communications Quarterly* 15, no. 3 (2002): 410–438, doi: 10.1177/0893318902153004; Laurie A. Rudman, "Self-Promotion as a Risk Factor for Women: The Costs and Benefits of Counterstereotypical Impression Management," *Journal of Personality and Social Psychology* 74, no. 3 (1998): 629–645, http://www.ncbi.nlm.nih.gov/pubmed/9523410; Corinne A. Moss-Racusin, "Disruptions in Women's Self-Promotion: The Backlash Avoidance Model," *Psychology of Women Quarterly* 34, no. 2 (2010): 186–202, doi: 10.1111/j.1471-6402.2010.01561.x.

6. What's the difference between bragging and positive disclosure? For a linguistic/rhetorical exploration of self-praise on Twitter, see Daria Dayter, "Self-Praise in Microblogging," *Journal of Pragmatics* 61 (January 2014): 91–102, doi: 10.1016/j.pragma.2013.11.021.

7. Jonah Berger and Katherine L. Milkman, "What Makes Online Content Viral?," *Journal of Marketing Research* 49, no. 2 (2012): 192–205, http://journals.ama.org/doi/abs/10.1509/jmr.10.0353.

8. Robinson Meyer, "Why Are Upworthy Headlines Suddenly Everywhere?," *Atlantic,* December 8, 2013, http://www.theatlantic.com/technology/archive/2013/12/why-are-upworthy-headlines-suddenly-everywhere/282048; Michael Reid Roberts, "Life Sentences: The Grammar of Clickbait," *American Reader,* http://theamerican-reader.com/life-sentences-the-grammar-of-clickbait; Belle Beth Cooper, "A Scientific Guide to Posting Tweets, Facebook Posts, Emails, and Blog Posts at the Best Time," *BufferSocial* (blog), Buffer, August 29, 2013, https://blog.bufferapp.com/best-time-to-tweet-post-to-facebook-send-emails-publish-blogposts.

9. Christopher Mims, "Why You'll Share This Story: The New Science of Memes," *Quartz,* June 28, 2013, http://qz.com/98677/why-youll-share-this-story-the-new-science-of-memes.

10. Alexia Tsotsis, "Google+ Is Walking Dead," *TechCrunch,* April 24, 2014, http://techcrunch.com/2014/04/24/google-is-walking-dead; Adrienne La France, Adrienne Meyer, and Robinson Meyer, "A Eulogy for Twitter," *Atlantic,* April 30, 2014, http://www.theatlantic.com/technology/archive/2014/04/a-eulogy-for-twitter/361339.

11. Kyle Vanhermert, "The Design Thinking behind Twitter's Revamped Profiles," *Wired,* April 9, 2014, http://www.wired.com/2014/04/the-design-process-behind-twitters-revamped-profiles; Dante D'Orazio, "YouTube Changes Search to Rank Based on How Long Users Watch Videos, Not Clicks," *The Verge,* October 12, 2012, http://www.theverge.com/2012/10/12/3494826/youtube-changes-search-to-rank-based-on-how-long-users-watch-videos; Kurt Wagner, "Facebook's News Feed: What Changed and Why," *Mashable,* April 16, 2014, http://mashable.com/2014/04/16/news-feed-changes.

12. Clay Shirky, "Newspapers and Thinking the Unthinkable," *Shirky* (blog), March 13, 2009, http://www.shirky.com/weblog/2009/03/newspapers-and-thinking-the-unthinkable.

13. Maureen Morrison and Abbey Klaassen, "Dan Wieden's Advice for Small Agencies: Learn to Fail," *Advertising Age,* July 25, 2013, http://adage.com/article/agency-news/dan-wieden-s-advice-small-agencies-learn-fail/243329.

14. Google Analytics, accessed February 14, 2015, http://www.google.com/analytics.

15. Jacob Smith, "Show Me Your Nonprofit Dashboard!," *Beth's Blog,* April 21, 2011, http://www.bethkanter.org/nonprofit-dashboard.

16. Shauna M. VanBogart, "Establishing Credibility Online through Impression Management," master's thesis, Gonzaga University, 2013, 47, http://pqdtopen.proquest.com/doc/1501649062.html?FMT=AI.

17. Jacquelyn Gill, post on Twitter, February 6, 2014, https://twitter.com/JacquelynGill/status/431453808674091008.

18. Nitsuh Abebe, "Watching Team Upworthy Work Is Enough to Make You a Cynic. Or Lose Your Cynicism. Both. Or Neither," *New York Magazine,* March 23, 2014, http://nymag.com/daily/intelligencer/2014/03/upworthy-team-explains-its-success.html.

19. Jeffery Pfeffer, Christina T. Fong, and Robert B. Cialdini, "Overcoming the Self-Promotion Dilemma: Interpersonal Attraction and Extra Help as a Consequence of Who Sings One's Praises," *Personality and Social Psychology Bulletin* 32, no. 10 (November 2006): 362–374, http://www.gsb.stanford.edu/faculty-research/publications/overcoming-self-promotion-dilemma-interpersonal-attraction-extra-help.

20. Dan Zarrella, "New Data Proves 'Please ReTweet' Generates 4x More ReTweets [Data]," *HubSpot* (blog), May 31, 2011, http://blog.hubspot.com/blog/tabid/6307/bid/14982/New-Data-Proves-Please-ReTweet-Generates-4x-More-ReTweets-Data.aspx.

21. Hauke Riesch and Jonathan Mendel, "Science Blogging: Networks, Boundaries and Limitations," *Science as Culture* 23, no. 1 (2014): 51–72, doi:10.1080/09505431.2013.801420.

22. Liz Neeley, "'So Tweet This, Maybe?'—Promoting Your Work in Social Media," *COMPASS Blogs,* February 25, 2013, http://compassblogs.org/blog/2013/02/25/so-tweet-this-maybe.

23. Liz Neeley, post on Twitter, January 22, 2014, https://twitter.com/LizNeeley/status/426232442089766912.

21

Blogging at Scientific Conferences

TRAVIS SAUNDERS AND PETER JANISZEWSKI

One of the biggest resources for news on the latest scientific discoveries is the scientific conference. Many of these conferences have become more open to sharing their results with science bloggers. How should you cover these conferences? What are the dos and don'ts? In this chapter, Peter Janiszewski and Travis Saunders share their experiences with blogging scientific conferences.

Academic conferences are a key conduit of information between scientific researchers. At conferences, hundreds or even thousands of researchers may gather in one location to share the latest research findings and insights on a given topic. While that may seem like a lot of people, it's easily overshadowed by the total number of individuals who may be interested in the information discussed, but who couldn't attend the conference in person—including nonattending researchers in that field, researchers from related fields, and laypeople with an interest in the topic. The largest conferences put out press releases of the most exciting research presented. Eager reporters covering the science beat might also make their way to a select session or two and report on the findings in a local paper or online. But by and large, the science that is presented at conferences stays at conferences.

Fortunately, this situation presents a fantastic opportunity for science bloggers who happen to be attending these conferences. One of the toughest things about writing a science blog is regularly coming up with timely content that has not already been discussed elsewhere. Conferences can serve up fresh and largely exclusive content for days on end. In addition, your blogging efforts can increase the exposure given to important research presented at the conference, or even the conference itself. This is a win-win-win for you, your colleagues, and the organization holding the conference. In this chapter we outline a variety of approaches to using blogs and social media during a scientific conference, and give some practical advice to help make your conference blogging a positive and worthwhile experience.

What Is Conference Blogging?

As far as we know the term "conference blogging" has never been strictly defined, and this seems like a good opportunity to do just that. Conference blogging refers to the use of any online media to share information from an academic conference or presentation with the rest of the world. This can be accomplished via a blog, a Tumblr, a Facebook page, a Twitter or LinkedIn account, a YouTube page, a podcast, or any combination of these. The point is to use social media to disseminate information beyond those physically present at a conference.

Before the Conference

First and foremost, be sure to contact the conference organizers beforehand to notify them that you are interested in blogging about the conference. Some conference organizations may not be supportive of blogging, while others may have specific rules regarding how this should be done (that is, no photos or audio recordings of presentations). In either case, it is best to acquire this information before attending the conference. Otherwise you risk getting escorted from a conference session for inappropriate behavior. Good luck explaining that to your colleagues!

Assuming that blogging is permitted, you may actually be able to get some level of support from the conference organizers. We've found that the level of support provided can vary substantially between conferences. Presumably, the prominence of your online platform will at least partly determine the level of support provided, but so too can the financial health of the conference organization, or the organizers' awareness and perception of social media. On the less impressive end of the spectrum you may be given access to free wifi, access to a media lounge (free donuts!), or early access to the conference program and abstracts. If you're more established as an online science communicator, you may be listed as an "official conference blogger" on the conference website or program, or receive free registration. And finally, if you're a blogosphere superstar, you could receive free travel and/or accommodations. The most important thing to remember is always to ask for support—it never hurts, and from our experience, conference organizers are generally open to providing some perks in exchange for the free publicity that blogging can provide.

Next, whenever possible, go through the conference program ahead of time and identify any key sessions that you think would be a good fit for your blog. This is also a good time to contact people whom you might want to interview during the conference. You will want to gauge whether these individuals are interested in discussing their work with you, and whether they'd prefer to do so in person or via email. If you're lucky, you might even get a few interviews under your belt before the conference officially starts, saving you lots of time at the conference itself.

Finally, before heading to the conference, consider putting up a blog post to let people know how to get in touch with you while you're there. You could also join or start an online pre-conference conversation with other attendees on Twitter. The easiest way to accomplish this is to begin using the official Twitter hashtag for the conference whenever discussing anything relevant. While most conferences today establish and widely advertise their official hashtags, you may still come across a small or digitally unsavvy conference without an official hashtag. In this case, do a quick search on Twitter to see if others in attendance have already started using an unofficial hashtag. If not, feel free to go out on a limb and make up your own (we suggest keeping it as short and distinctive as possible—the conference initials and year are a common method). Before you know it, the hashtag you invented may become the common link for all online conversation at the conference.

During the Conference

In our experience, it is *much* easier to blog about a conference while you are actually there, rather than when you get back home. Once the conference ends you'll quickly become too busy catching up on all that you missed while away to post much content. (Travis has several audio interviews from a conference two years ago that he still hasn't gotten around to publishing.)

Keep in mind that blogging a conference can become overwhelming if you aim to produce more content than is reasonable. Make sure to set a realistic estimate of how much time you can devote each day to blogging, then stick to it as best as possible. Although everyone has a unique approach, a great strategy we've adopted is to dedicate a chunk of time during each day to post content. This ensures we're not stuck blogging in the wee hours of the night and missing out on social events. For example, Travis has found that just thirty to sixty minutes each day provides enough time to post content, and can often be squeezed in between conference sessions. Since you can't be everywhere at once, you can also ask other conference attendees to write guest posts about sessions that you did not attend. Whatever approach works best for you, just remember that blogging should supplement but not take over your entire conference experience.

As soon as you put up a blog post, send a brief note to anyone mentioned to alert them that the content is going live and to thank them for their contributions. People always appreciate this little courtesy. And because most of us are suckers for seeing our name in print or our face in a photo or video, the people mentioned in your post are likely to share the content with their friends, family, and colleagues, bringing new visitors to your site.

Finally, try and stay abreast of the online dialogue at the conference. You can set up a search for the conference's Twitter hashtag to see what the other attendees are discussing online. This can become particularly useful when multiple attendees listening to the same presentation discuss the content presented in real time. Setting up a Google Alert (see http://www.google.com/alerts) for the conference can also help you find online coverage of the conference.

After the Conference

Once the conference is over, take a look at your analytics to get an idea of how many people viewed your content through your different online platforms. Tweetreach (http://tweetreach.com) is a useful tool for tracking the number of tweets using a specific conference hashtag, and Google Analytics (https://www.google.com/analytics) is excellent for assessing traffic to your site. Be sure to pass all this information along to conference organizers and anyone whose work you featured during the conference. Conference organizers and researchers really appreciate this sort of quantitative information, and keeping them happy may lead to increased acceptance of social media in the conference setting—and more opportunities for you.

Essential Tools for Conference Blogging

Before you set off to blog your first conference, it is essential to become comfortable with a number of tools that will make this possible. In terms of hardware, you will need at least one of the following items: (1) a compact laptop or netbook with good battery life, (2) a tablet, or (3) a smart phone. Now let's take a look at some of the most common online tools for disseminating conference news, from least to most time consuming.

TWITTER

By far the simplest approach to conference blogging (technically, microblogging) is to simply tweet about interesting things that you've come across at a conference in 140 characters or less. Also, you can use Twitter to link to other content related to the conference, including your own or others' blog posts, videos, slideshare presentations, and more.

Many organizations now explicitly encourage Twitter use by promoting conference-specific hashtags. By setting up a Twitter search for that hashtag you can easily follow and participate in the online discussion taking place among people at the conference. To see how this works in practice, go to Twitter and search for #AHKCSummit; the official hashtag of the 2014 Global Summit on the Physical Activity of Children. You will see links to newsworthy presentations, photos of posters that were presented at the conference, updates from conference staff on what is happening in each session at a given time, links to blog posts that give extra detail on topics that were discussed at the conference, and pictures from social events.[1] Conferences like the Global Summit are a great example of the way that a professional gathering can use social media (and Twitter especially) to transcend its physical location—tweets using the hashtag begin to pop up well before the conference each year, and continue for months after people have gone home.

Live-tweeting a conference takes very little time or effort, which makes it an ideal way to begin conference blogging. Even if you tweet only infrequently, you can keep up with the latest happenings simply by following the conference hashtag online. This is especially useful at large conferences with multiple sessions occurring simultaneously. Twitter can also serve as a powerful networking tool while at a conference. If, like us, you are a tad socially awkward, it can be much easier to introduce yourself to another researcher in person if you've previously participated in some scientific repartee via Twitter. If you are looking for an easy way to get into conference blogging, then Twitter is a great place to start.

A useful tool that helps leverage the power of Twitter is Storify, which allows you to organize tweets on a specific topic (for example, by using a conference hashtag). Storify also allows you to insert your own commentary between tweets, and can be embedded into blog posts. For example, *F1000* used Storify to present tweets from a session on science publishing at Experimental Biology 2013, then embedded that Storify within a blog post that provided additional information and context.[2] This can be a great tool for providing people with an overview of a presentation using tweets from those in attendance, and for exposing/repurposing a Twitter conversation for a broader audience.

PHOTO SHARING

People like pictures! Just look at the popularity of Instagram, and the volume of photos shared on Facebook. So if you are blogging or tweeting at a conference, make sure that you snap a few pictures to post online. Photos can supplement your other content and help to attract people to posts that they might otherwise skip, or they can stand alone. Either way, photos can be a great way to network with other attendees. A great example of this approach comes courtesy of Carin Bondar, who took pictures with nearly everyone attending the ScienceOnline conference in 2011. She then posted many of these pictures online with a tag identifying the other attendees. Taking and sharing photos is a wonderful way to introduce yourself to countless people, and to give a sense of what is happening at the conference to those following from afar.

VIDEOS AND PODCASTS

Given the ubiquity of cameras and voice recorders in smart phones, it couldn't be simpler to record audio and video clips and publish them online while attending a conference. By live-streaming a presentation (which is possible using YouTube, UStream.tv, Justin.tv, Google+, and other tools), you can even allow people to watch a conference talk in real time. In fact, when combined with a Twitter hashtag, this approach can allow online viewers to pose questions to the presenters. Just be aware that many conferences have very specific rules regarding this type of activity—photos, and audio and video recordings, may be restricted during some sessions. Also, since information presented at a conference may not have been officially published, the presenter may not appreciate your posting photos of his or her slides on the web. Always ask the conference organizers as well as the individual presenters if they're open to having you share their presented materials online.

Another simple way to create content is to conduct short interviews with other conference attendees. People almost always say yes to an interview, and a three- to four-minute video or audio recording can be uploaded to YouTube or a podcast hosting site such as Podomatic.com almost immediately.[3] The speed of this approach is a major bonus for you as a conference attendee, because it allows you to disseminate content without taking much time away from your own conference activities. Video and audio are also extremely easy to share further, since people can embed videos and podcasts on their own blogs and Facebook pages. Embedding video and audio files in this way can help them to spread incredibly quickly.

BLOG POSTS

The most obvious way to "blog" about a conference is through a traditional blog post. Blogging can take the form of interviews, descriptions of interesting presentations, or your thoughts on a controversy that erupted during the conference. Unless you're a particularly expeditious writer, however, be aware that writing long-form blog posts can take a lot of time away from your conference experience.[4] We have found that alternating moderate-length text-based posts with shorter video- or audio-based posts can be a very good way to publish a lot of content while at a conference without overwhelming yourself.

Beware the Dreaded Ingelfinger

The Ingelfinger rule is named after former *New England Journal of Medicine* editor Franz Ingelfinger, and states that a manuscript will not be accepted for publication in a journal if it has already been published somewhere else.[5] Technically, this means that if you publish the results of a study on your blog, it may subsequently be rejected from publication in a peer-reviewed journal. This is obviously a concern for researchers. So although we have yet to hear of any instance of this rule being applied in practice, it may be best to proceed with caution when discussing results that have yet to be published in a peer-reviewed journal (read: almost all original research presented at conferences).

It is critical to get permission from any researcher before discussing his or her original unpublished research on your blog, and especially before posting any images of posters, tables, or figures. Discussing research that has already been published, which is often the case in keynotes and plenary sessions, is much less of a concern. When reporting on unpublished data, you might want to leave out details (exact numbers, statistics, and so on), and stick to the general findings of the study. Of course, it is common for abstracts of conference presentations to be available online or to be published in a journal supplement—sometimes before the scheduled conference. In our opinion, the information in the abstract is therefore fair game for use in a blog post or other online discussion.

If you are also presenting at the conference, you may want to avoid discussing your own discoveries if they have yet to be published. As vice president and global editorial director at *MedPage Today,* Ivan Oransky, explained in an interview with the blog *Scholarly Kitchen:* "Ingelfinger . . . at least officially, doesn't prohibit prepublication publicity as long as researchers don't court journalists' attention."[6] In our interpretation, this means that, with another researcher's permission, it's kosher to blog about his or her unpublished research, but not to blog about your own unpublished research (yes, it seems strange to us, too).

At the time of this writing, there exist no official rules pertaining to conference blogging. We hope that the advice we've provided in this chapter will help ensure a positive experience for the aspiring conference blogger. The most important thing is to make sure that the conference has no specific edicts forbidding blogging, and then to decide how you can best contribute to the online discussion using the tools described here. And remember, conference blogging should be fun—if you're getting overwhelmed by the amount of work that blogging adds to your conference schedule, consider scaling it back a notch.

A Summary of Conference Blogging Dos . . .

- Contact the conference organizers and potential interviewees ahead of time
- Take plenty of notes on topics that might lend themselves to a blog post
- Follow the conference hashtag on Twitter
- Create a Google Alert for the conference
- Link to or embed other conference content that you come across online
- Invite other conference attendees to write guest posts
- Ask researchers' permission to discuss their unpublished research

. . . and Don'ts

- Let blogging take over the whole conference for you

- Take a picture or other recording during sessions when such recordings are prohibited
- Post pictures of figures or other copyrighted material without permission
- Ingelfinger yourself

TRAVIS SAUNDERS is an assistant professor in the Department of Applied Human Sciences at the University of Prince Edward Island. **PETER JANISZEWSKI** is a writer, researcher, and blogger. They are the co-founders of the health blog *Obesity Panacea* on *PLOS Blogs*, as well as the *Science of Blogging*, a blog focusing on ways that social media can be used in science communication.

Travis is based in Prince Edward Island, Canada. Peter is based in Toronto. Find them at their websites, http://www.upei.ca/science/travis-saunders and http://peterjaniszewski.com, or follow them on Twitter, @TravisSaunders and @Dr_Janis.

Notes

1. AB Ctr 4ActiveLiving, post to Twitter, May 22, 2014, https://twitter.com/4ActiveLiving/status/469499902796763136; Tim Takken, Ph.D., post to Twitter, May 22, 2014, https://twitter.com/bikedocter/status/469428314185007107; Lizz Picooli, post to Twitter, May 22, 2014, https://twitter.com/piclizz/status/469491430856355840; Travis Saunders, post to Twitter, May 21, 2014, https://twitter.com/TravisSaunders/status/469228905715335168; Deirdre M. Harrington, post to Twitter, May 22, 2014, https://twitter.com/DeeHarrPhD/status/469481056031088644.

2. Eva Amsen, "Storify of 'Challenging the Science Publishing Status Quo,'" *F1000 Research* (blog), April 25, 2013, http://blog.f1000research.com/2013/04/25/storify-of-challenging-the-science-publishing-status-quo.

3. Travis Saunders, "Swap Sitting for Sleep to Improve Your Health?," *Obesity Panacea* (blog), Public Library of Science, November 2, 2012, http://blogs.plos.org/obesitypanacea/2012/11/02/swap-sitting-for-sleep-to-improve-your-health-icpaph12-beactive2012.

4. Scicurious, "How to Blog a Conference," *Science of Blogging,* http://scienceofblogging.com/guest-post-how-to-blog-a-conference.

5. L. K. Altman, "The Ingelfinger Rule, Embargoes, and Journal Peer Review," *Lancet* 347, no. 9012 (1996): 1382–1386.

6. Kent Anderson, "Interview with Ivan Oransky of *Retraction Watch,*" *Scholarly Kitchen,* August 3, 2012, http://scholarlykitchen.sspnet.org/2012/08/03/interview-with-ivan-oransky-of-retraction-watch.

22

Tackling the Hard Sciences

RHETT ALLAIN

Tackling the "hard" science subjects like math, physics, and engineering in one's writing can be a daunting task. The Internet, where misinformation abounds and critics are eager to advance their pet theories, adds an extra twist. Rhett Allain will explore the different ways of writing about the hard sciences on the Internet, and explain how the web can provide additional context to the findings of everything from math to astronomy.

In many ways, blogging about "hard sciences" like physics is just like blogging about other, "squishier" areas of science. There are generally four different types of science blog posts: reporting, explaining, analysis, and link filtering. Of these, posts that report on an event while explaining the physics (or other scientific principles) involved are the best, in my opinion. For examples of fine blogs that have more of the "reporting"-type posts, take a look at *Physics Buzz* (http://physicsbuzz.physicscentral.com) or *Bad Astronomy* (http://www.slate.com/blogs/bad_astronomy.html).

Reporting

What kinds of events would a physics blogger report and write about? Some examples might be the 2013 Russian meteor event, the 2012 Higgs boson detection, or the 2011 faster-than-light neutrinos anomaly. These are events that are reported all over the place. So what makes science bloggers different from other news sources? Even if they don't offer an explanation to go along with the event (though they usually do), they have a level of expertise that allows them to point out the important aspects as well as dispel any incorrect information. They also have the advantage of freedom. A blog post can report the news and include analysis in whatever way the blogger chooses.

Here are a couple of examples of "reporting" from my own blog (http://www.wired.com/wiredscience/dotphysics). Note that you could probably consider these posts as fitting into more than one category.

- What Can We Do with the Higgs Boson?: This post doesn't report the details of the discovery of the Higgs boson, but it is still a type of reporting (http://www.wired.com/2012/07/what-can-we-do-with-the-higgs-boson).
- Watch the Red Bull Stratos Jump Live: This is just reporting. In this post, I simply share details about how you could watch the Stratos Jump, a skydive from 120,000 feet (http://www.wired.com/2012/10/watch-the-red-bull-stratos-jump-live).

- Where Does Carbon Come From?: The intention of this post was to report on verification of the nuclear process that leads to the creation of carbon. It does that, but I first explained how we get carbon (http://www.wired.com/2011/05/where-does-the-carbon-come-from).

Explaining

At its most basic level, this kind of post is just an explanation. The blogger could be describing something simple, like the different colors of light or the nature of force and motion. Or she could be tackling something very complex such as a numerical solution to the energy levels in a hydrogen atom. I like to think of these types of posts as textbooks without constraints. The goal is to provide some enlightenment on a concept, and there are no limits to how simple, how complex, or how long the explanation can be. You also don't have to assign homework (but you can). A blog can contain links to other material, animations and video, as well as various explanations aimed at readers with different levels of background knowledge about the subject. Sometimes my posts are about advanced physics concepts and other times they are very basic. Most textbooks can't do that.

Just about every science blog offers this type of explaining at some level. Personally, it is one of my favorite types of post. Examples from my own blog include a post on Gollum (from the book and movie versions of *The Hobbit*) to explain how humans see and how we could possibly see in the dark.[1] And in "Why Do Astronauts Float around in Space?" I just answer the question posed in the title of the post.[2] It's all just explanation.

Are there things to explain in other disciplines? Of course. Why are some bacteria resistant to antibiotics? How does a computer work? Explanations are needed in every field.

Analysis

What is an analysis post? Let me start to answer that question with an example of one of my favorite types of analysis posts: video analysis. In video analysis, I start with a piece of video and find data to share about that video. For instance, suppose there is a crazy video of an eagle picking up a child.[3] Is this real or fake? Answering this question would be the main point of one of my video analysis posts. Another popular class of analysis posts is the "what if" type. What if everyone on the Earth jumped at the same time? What if I wanted to start a car with D-cell batteries? How is an analysis different from an explanation post? In an analysis post, you are trying to show some possibly new thing (even if that thing actually isn't new). It usually involves some type of calculation or data collection. An explanation post doesn't require any data or calculations.

A speed test between the DC Comics superhero Flash and the Marvel Comics mutant Quicksilver is very typical of my analysis posts.[4] I like to take something that could be real or is clearly not (it doesn't matter) and then apply physics models to the situation. Another analysis post examines the depiction of gravity in the iPhone game Angry Birds Space.[5] Do you think a peer-reviewed journal would accept a publication about superheroes or Angry Birds? Probably not.

Link Gathering

How is any mere mortal able to sift through all of the eye-catching posts online and find the good ones? It seems like an impossible task. This is where your blog can come to the rescue. The best blogs take the time to filter through the massive Internet and find the best pieces of science for their readers to consume. Consider *Swans on Tea* (http://blogs.scienceforums.net/swansont) or *Physics and Physicists* (http://physicsandphysicists.blogspot.com). Perhaps the best example is Jennifer Ouellette's *Cocktail Party Physics* and her "Physics Week in Review" (http://blogs.scientificamerican.com/cocktail-party-physics). All of these do an excellent job of finding and sharing great links—connecting readers to up-to-the-minute, accurate material in ways that traditional media can't. I don't post links too often on my own blog, but linking is the peer-review process of the Internet. Things with more links could be seen as being more valuable (although this is not always true).

Beyond the Text

Where do you get ideas for blog posts? If you're like me, you answer questions that you have yourself. When I see something in the news that looks interesting and physics-related—which isn't difficult given that just about everything that moves or is powered by electricity is connected to a physics concept—I will make that a blog post. Sometimes, too, it will be a question that someone else asked. For instance, someone asked why the moon doesn't crash into the Earth even though the Earth pulls on the moon.[6] That's a great question. It is fun to try and explain these things in a way that large numbers of people can understand.

There are two other great sources for physics blog posts. The first is movies. The physics of movies is usually wrong, making them a great source for posts. I think I could take just about any movie and make a physics post about it. If you want to connect with your readers, who may or may not be physics experts, movies can be an excellent way to go. Each scientific discipline has its own connections to the real (or fictional) world. This is what makes it so much fun to be an expert in some particular area.

The other very useful source for blog ideas is the Internet itself. Just start wandering around and see what questions are being asked.

My own favorite sources for ideas are the social networking sites Reddit and Twitter. While on these sites, I look for anything that can spark an idea. It could be an awesome video, or maybe an unsettled argument. All you have to do is pick the topic and find out what scientific concept applies. It's like an instant blog post right there. As a bonus, it is probably a video or discussion that is already popular. By adding the analysis or explanation, you are contributing to the usefulness of these Internet ideas.

Something else to consider for your blog posts: who is your audience? Sometimes you aren't sure. A blog isn't like a lecture or a presentation; instead, it is like a billboard on the side of the road. You might have an idea of who will pass by and see it, but you don't know for sure. This can make it difficult to choose a writing level. Should you assume that your readers already understand Newtonian mechanics or should you start from scratch? Really, it's a decision you need to make for each post. One of the nicest things about a blog is that you can write at whatever level you like, and unlike a magazine or newspaper article, you aren't limited by length. Sometimes it is very helpful and interesting to provide every single detail needed to support a concept or idea. It may be one of the greatest things a physics blog can do—to show all the details that are needed.

What about blogging tools? This might be an area where "hard science" blogs differ most from other blogs. I can't imagine a physics, math, or engineering blogger who would never use graphs. Scientists in general love graphs, but often physicists need them. How can you share the trajectory of an object without a graph (unless you show an animation)? There are several ways you can create a graph. Pick one and become an expert. This way you will know all the tricks and can work quickly to put your graph online. Some of the options include:

- Microsoft Excel. Excel might offer the simplest way to make a graph, particularly if you are already familiar with Microsoft Office products. But I rarely use it. Why? First, I think it was made with business people in mind and not scientists. Just look at the method for fitting a function to data: it is called the "trendline." You can't get more businesslike than that. The other problem I have with Excel graphs is that I find them less than pretty. Sure, beauty is in the eye of the beholder, but I'm just not fond of the way these graphs turn out. There are other software packages for graphing that are similar to Excel: Open Office, Google Docs, Apple's Numbers. I would group these all together.
- Python. What do I use to make graphs? I almost always use the high-level programming language Python. Python is very simple to learn, and what makes it awesome are the modules—groupings of related, ready-made lines of code. You want to make a graph? Just import a graphing module (like Matplotlib). For me, this strategy has a number of advantages. First, the results are pretty. Second, since it is a module in Python I can plot things as a result of a calculation. It might take some time to learn a nice graphing tool like Matplotlib, but doing so will save time in the long run.
- Plotly. Another graphing option is the online tool Plotly (http://plot.ly). This free tool allows you to make both simple and complex graphs without too much fuss. What's even better is that you can share them online.

What about equations? You might be tempted to just write out in your text an equation such as $x = (1/2)at^2$ to represent that distance is equal to one-half the acceleration multiplied by the time squared. This does get the point across and it is simple. But what if you want to do something more complicated? What if you want to show an integral? Here is where you will need some method for making equations. LaTeX is a type of document formatting code—a typesetting programming language designed for equation-heavy documents. There is a LaTeX plugin for WordPress that will allow you to write LaTeX code in your blog and, just like magic, it will appear as a nice equation. You could think of it as the paper version of HTML in that you don't see what your paper looks like until you render it.

If you don't have a plugin, you could use some other method for creating your equations. I like to use a LaTeX equation editor for Mac OS X, but you could also use Microsoft Equation Editor or an online equation editor (see http://www.numberempire.com/tex equationeditor/equationeditor.php). From here, you can save your equation as an image and then insert it into your blog. Just like the graphs, I would say it is important to become proficient with your tool of choice.

There is one more tool that I can't get by without—something to make illustrations. Just about every science blogger at some point is going to need to make a drawing of something. Maybe it's a diagram to show forces or something to indicate how dust interacts with air. Sure, you could do a Google image search to find a suitable diagram—but you might not find exactly what you want, or maybe it isn't free to use. It's always better to make your own. How do you make diagrams? Whatever you use, you want to be an expert. Just about all of my drawings are created in Apple's Keynote software. Yes, I know that is meant for presentations and not really for drawings, but I can make just about anything with the drawing tools, so that's what I use. And about pictures: you have a phone, right? It probably has a camera. Take pictures of things like bananas. You never know when you will need a banana picture. Again, it is better to use your own since it is exactly what you want. For moving images, I also like to use the free video analysis tool Tracker Video (https://www.cabrillo.edu/~dbrown/tracker). This lets me look at the positions of objects in each frame of a video. It is very useful when writing about YouTube videos and movies. The other thing that I use quite often is, again, the programming language Python, because there are so many situations that can be modeled with numerical calculations. If you don't like Python, you could use any number of other packages. Again, pick your tool and become an expert.

Before you start your own blog in the hard sciences, then, try to think of the type of posts you might like to write and consider which tools you will want to use. Remember what Abraham Lincoln said about science blogs: "Give me six hours to chop down a tree and I will spend the first four sharpening the axe." If you take the time to master your tools and sharpen your skills, when the time comes to get a quick turnaround on a hot topic, your blog post will be ready.

RHETT ALLAIN is a physics professor at Southeastern Louisiana University. He also writes the blog *Dot Physics* at *Wired*, and is the author of the book *Geek Physics*.

Rhett is based in Hammond, Louisiana. Find him at his blog, http://www.wired.com/category/science-blogs/dotphysics/, or follow him on Twitter, @rjallain.

Notes

1. Rhett Allain, "How Does Gollum See in the Dark?," *Wired*, December 10, 2012, http://www.wired.com/2012/12/how-does-gollum-see-in-the-dark.

2. Rhett Allain, "Why Do Astronauts Float around in Space?," *Wired*, July 9, 2011, http://www.wired.com/2011/07/why-do-astronauts-float-around-in-space.

3. Rhett Allain, "Eagle Picks Up a Kid: Real or Fake?," *Wired*, December 19, 2011, http://www.wired.com/wiredscience/2012/12/eagle-picks-up-a-kid-real-or-fake.

4. Rhett Allain, "Who's Faster? Flash or Quicksilver?," *Wired*, June 5, 2014, http://www.wired.com/2014/06/whos-faster-flash-or-quicksilver.

5. Rhett Allain, "The Gravitational Force in Angry Birds Space," *Wired,* March 29, 2012, http://www.wired.com/2012/03/the-gravitational-force-in-angry-birds-space.

6. Rhett Allain, "Why Doesn't the Moon Crash into the Earth?," *Wired,* November 29, 2012, http://www.wired.com/2012/11/why-doesnt-the-moon-crash-into-the-earth.

23

Blogging about Controversial Topics

EMILY WILLINGHAM

Climate change. Vaccines and autism. Creationism. The scientific world is full of topics that incite anger and attract armies of trolls. Freelance writer Emily Willingham is no stranger to the deluge that can follow when blogging about controversial topics. In this chapter, she will discuss her experiences and prepare you to blog about the things everyone loves to hate.

I don't blog about controversies to be controversial. I don't do it for the clicks. I do it because I am passionate about my primary subject area, autism, and I have strong opinions about the scientific, public health, and social issues associated with autism misinformation. The work of blogging about the facts is critical on a personal level because I have an autistic son, and a world that views him as a vaccine-injured, "toxic" mutant who needs to be fixed is a world I want to change for him.

A Passion and an Investment

I can't recommend purposely blogging about controversial subjects, but the fact is, the more controversial your subject area, the more eyeballs and clicks and link sharing you will get when you write about it. Autism, which as a human neurobiological condition might not seem terribly controversial at first glance, is one of the most polarizing subjects in our culture today. I also write about other controversial issues in science, but autism encompasses debates related to the environment, public health, disability rights, and parenting, all minefields of righteousness and judgmentalism. And boy do people like to fight. When I write something straightforward and terribly sincere about autism, I'll get a few thousand reads, maybe a handful of comments. But posts about vaccines and autism? I've had days I've thought that all I need to do is write "vaccines autism vaccines autism vaccines autism" about a hundred times, and I'd get the most readers I've ever attracted with a single post.

Passion and investment are critical if you're going to blog about controversies. You have to be consistent in your approach to the material and willing to alter your conclusions as new data come in. When you build trust in this way with the people who read your work, they will come to you for clarity when they read something somewhere else that leaves them scratching their heads. Responsibility and transparency also are paramount when controversy is your beat because even a hint of a conflict of interest or a failure to be intellectually honest about your material can torpedo trust. And it can harm your goals. I blog precisely because I think that the subject that is controversial shouldn't be, and I want to cut through the noise and get to the heart of what's factual about it, minus the muppet flailing.

Some people who blog about controversial subjects do so for venal purposes. Whether we like it or not, writers occupy a world that focuses more and more on personal "branding"—getting clicks and ad revenue—than on great reporting and analysis. For some people, writing for clicks and social media shares works. It doesn't work for me. Money is great, but it's not what motivates my blogging. My motivators are my passion for my subjects and my obsession with straight talk.

The drawback of starting from passion and a belief in the effectiveness of straight, honest discussion is how vulnerable it can leave the writer. So when you bring passion to your writing, you also have to bring your thickest skin. Or at least your thickest online skin costume. What you do where no one can see you, how you react in real life, in the glow of your laptop? That's your business. But online, one key to maintaining reader trust is to stay poised. And that's the hardest part.

How do you go about writing with poise about a controversial subject that impassions you? First, make sure to write without using a lot of emotion words—or, as one of my commenters once called it, "emotional valence." Emotion has its place, and I do express frustration, annoyance, disdain, happiness, and other emotions at reasonable levels. But I work very hard to avoid letting emotion drive or support my arguments. I don't base what I write on feelings. If you're going to write about controversial topics, you'd better line up your data ducks—facts swimming along in accessible, flowing prose—and leave the feelings for those who react to your words.

Another key is to avoid making it personal, both in what you write originally or in your response to comments. I know that some bloggers have built entire communities around calling people "idiot," "moron," and other epithets suggesting the intellectual inferiority of those who don't agree with them. But I think that engaging in this kind of attack-the-person tactic simply causes anyone on the fence to withdraw from your arguments to avoid the blast of your epithets. If you're writing about a controversy just to get your choir to come in and sing your praises, then firing away at opponents' personal characteristics will work just fine. But if your goal is to make compelling, evidence-based arguments that reach information seekers rather than true believers, then why alienate those sincere knowledge seekers by acting that way?

The Power of Linking

That takes me to the evidence part. In addition to avoiding emotion-laden phrases like "I feel" in support of my arguments, I also try to steer clear of "I believe." I let the evidence speak for itself. In fact, a frequent phrase I use when responding to comments that nitpick evidence is "My readers can look at the evidence I linked for themselves and draw their own conclusions." The linking is critical, and the choice of links is crucial. Take advantage of the fact that this is an online world and link any assertion you make to the supporting data or documentation. The best kinds of documentation for a scientific controversy obviously will be the science itself, so wherever possible, I link to original studies or comprehensive reviews, preferably those that are open access, so that people can look at the data themselves. I rarely rely on links to opinions or other blogs unless the people writing those are considered trustworthy and extremely capable analysts and experts in their fields. Trust is a two-way street, and I look for the same qualities in trusted sources that I require of myself.

The urgency of appropriate linking came home to me as I was writing this chapter, in a situation that reveals one of the potential dangers of blogging about controversial topics. I published a post on an analysis in the journal *Pediatrics* regarding the possible links between gut disruptions and autism. The *Pediatrics* authors made a point that they'd made previously: if it hadn't been for a key figure in this field being found guilty of misconduct and dishonesty in a very high-profile investigation of an autism-gut link, scientists might not have treated the subject like an untouchable hot potato. Because of the taint of the original controversy, investigations of gut problems in autism have not been as frequent as they might have been, according to these authors—and I agreed. In my post, I referred to the actions of the key figure in question as "fraud," but when I published the post I did not link that word to the abundant documentation available to support its use.

The upshot was that the key figure in question sent a letter threatening to sue me for libel. It's not a tenable threat because he is a public figure and he'd have to prove that my goal was to defame him, which it was not. But if I had only linked to the literally dozens and dozens of online articles and papers using exactly that term in the same context, that word wouldn't have looked like it was only mine; indeed, the editors of the journal in which his original, now-retracted paper appeared had themselves published an editorial using the word "fraud." But as I've said, it's an online world. I immediately added a link to one of the dozens of possible pages of online evidence of the validity of the term and placed a parenthetical note establishing that the addition was an update. That approach is also key to retaining trust. Don't ever alter something fundamental about a live post without making it clear that you've done so, or you will open yourself to accusations of sneakiness, at best—and there goes your reputation for being trustworthy. For most bloggers, the hosting service or publisher offers no protection from libel suits, so you should always support any potentially upsetting claims with clear evidence.

Comments

Obviously, you have control over what you write. But you can't say the same about those who post responses to your blog entries. You have no control over what they write or how they write it except maybe, depending on your publisher, by hitting the "delete" or "spam" button. Your commenters can engage willy-nilly in all kinds of personal attacks and diversions and throw up every kind of fallacy that exists (or even invent some new ones). Some outlets, such as *Popular Science*, have a rule about not allowing commenters at all. But if your publisher allows comments, research supports that engaging early with your commenters is key to ensuring that an appropriate tone prevails and that their responses don't negatively skew your readers' impressions about your article.[1] I keep this finding in mind when thinking about anyone who begins perusing comments on my posts, which can and do run into the hundreds.

My rules for managing comments are as follows:

- I don't respond to ad hominems, including demands to prove or establish my credentials or defend myself against false accusations of, say, being a "pharma whore." What I write stands or falls on the evidence, even though I'm heavily credentialed. If I'm subject to a particularly personal attack, I will borrow a line from Spock in the new Star Trek movie series: "Reverting to name-calling suggests you are defensive and therefore find my objections valid." I drop that line in and do not respond any further to that commenter.
- I respond to inappropriate language from commenters. I will call commenters on threatening language or language that is abusive, and on sites where I have control, I moderate comments closely and delete or block commenters who violate house rules.

- If I do respond, I keep my answers as brief as possible. Do not fall into the trap of repeating your points or arguments—it wastes your time and can be sustaining to egos that are already large enough. If your goal is to reach people on the fence who are bothering to read the comments, once is usually enough to make a point. In fact, don't engage in lengthy exchanges with any commenter unless it's a really interesting, revelatory discussion that touches on something unknown to you—in which case, now you've got more blog fodder.

In this age of multiple platforms and platform cross-pollination, your online presence will extend beyond your blog or other publishing tool. For anything I do in social media that is public—public posts on Facebook, Twitter exchanges—I try to follow the same rules of exchange I have for my blogging. The only exception is that I'm much faster to block people who clearly are engaging me just to try to get attention. I don't feed trolls or narcissists, regardless of the platform.

What's Next?

Speaking of blog fodder: what if your blog well runs dry? One good strategy is to establish yourself as a general controversialist. I write a lot about autism, but my wider "beat" is simply "debunkery" and bringing some common sense to a lot of the nonsense that gets disseminated as science. Second, use Google news alerts—which you can set up through your Google profile so that you receive an email when any term you specify appears on the Internet—and follow people who curate links and have discussions in your areas. That approach generally gives me plenty of fresh fodder; indeed, more than I can keep up with. Have an alert for people with whom you disagree—you'll be able to see their latest material very soon after it's published, which might give you some ideas. And finally, have a Google Alert on yourself, so you'll have a heads-up when someone blogs about *you*. The last thing you want is to wander over to Twitter some fine evening only to step unaware into a minefield of twenty tweets mentioning something ugly that someone's written about you.

Regardless of where you get your fodder, make sure that you also have the necessary energy level at hand when you hit "publish." You will have days when you have all the energy and time you need to deal with a deeply controversial part of your controversy, and your loins will be as girded as they can be. But there will also be days when one more straw might crack your brain—and those are the days to hold back on that controversial post. If you have no choice, for some reason, and have to press "publish," the world can wait if you need to take until the next day to open up that comment page and see what's happened in the interim. Sometimes that kind of waiting gives me the time I need to dial back my emotional reactivity to the subject. And if you wait, sometimes you might find yourself pleasantly surprised to see that others have come in and provided lucid, useful responses so you don't have to. Sometimes.

Finally, I have no way of predicting how successful a post will be, if you measure success in terms of attention relative to time invested. Posts I've spent careful hours on to plug every possible logic hole and avoid stirring up even more controversy can land with a thud of just a few thousand views. But in those cases I don't spend a lot of time on comments because, well, there just aren't as many. And posts I've spent just twenty minutes on, that practically wrote themselves in my head before I ever put fingers to keyboard, will catch like a viral inflammation through social media and blow up within an hour. The trade-off with those fast and fast-flying posts can be more extensive investment in comment responses, which sometimes continue for weeks. Either way, when you're blogging about controversies, there's no such thing as a "quick writeup." On the controversy beat, hitting "publish" on a blog post usually isn't the end. In most cases, it's just the beginning.

EMILY WILLINGHAM is a medical writer and freelance writer. She blogged at *Forbes*, and is co-authoring a book on evidence-based parenting scheduled for publication in 2016 by Perigee Books/Penguin.

Emily is based in San Francisco. Find her at her website, http://www.emilywillinghamphd.com, or follow her on Twitter, @ejwillingham.

Note

1. "The 'Nasty Effect': How Comments Color Comprehension," NPR, March 11, 2013, http://www.npr.org/2013/03/11/174027294/the-nasty-effect-how-comments-color-comprehension.

24

Persuading the Unpersuadable

Deniers, Cynics, and Trolls

MELANIE TANNENBAUM

A major difference between science blogging and traditional science writing is the presence of a "comments" section in blogs. In this chapter, Melanie Tannenbaum, psychologist and blogger at Scientific American Blog Network, *offers advice on how to maintain a civil and productive comments section, and discusses psychology-based strategies for persuading—and calming—your more quarrelsome audience members.*

Ah, the comments section—either the most loved or most hated part of a blog, depending on whom you ask. Comments can offer opportunities for a fulfilling discussion, or they can be cesspools of spam, trolls, and science deniers. In this chapter, I will suggest some ways to maintain a respectful comments section, and point to some strategies from the annals of psychological science that may help to get your more disagreeable readers on board. Science bloggers, after all, ought to take a scientific approach to managing their comment streams.

Respectful Commenting

One of the biggest ways that the people around us exert influence on our behavior is through the use of "norms," messages about what constitutes acceptable and appropriate behavior in a given setting. There are two main types of norms: descriptive and prescriptive. Descriptive norms simply describe the way that things are, whereas prescriptive norms offer a mandate about how things should be. For example, if I said that most college students go to class wearing jeans and sweatshirts, that would be a descriptive norm. If I said that you should wear jeans and a sweatshirt in order to fit in, that would be prescriptive.

Descriptive norms can be incredibly powerful. For example, in a classic study, Robert Cialdini and colleagues manipulated the signs that were displayed in Arizona's Petrified Forest National Park, a site plagued by tourists who would take fragments of petrified wood home as souvenirs.[1] In situations like this, the first inclination of well-meaning environmentalists might be to set a strong prescriptive norm, perhaps by writing on the sign something like, "Many past visitors have removed the petrified wood from the park, changing the state of the Petrified Forest. This is bad, don't do this." The idea here would be to invoke a sense of shame and severity before asking visitors to refrain from taking the wood.

But read that prescriptive message once again. That message is not just telling you that you shouldn't take the wood—it's also telling you that most other people do. In fact, people were actually *more* likely to steal wood from the forest when they saw the sign telling them how many people tend to do so. But when the researchers made the message read "The vast majority of past visitors have left the petrified wood in the park, helping to preserve the natural state of the Petrified Forest," the thievery plummeted.

We don't care so much about what we should do. We care about what other people do. And then we really, really care about not being different.

When bloggers think about how to moderate comments sections, they often instinctively try to set prescriptive norms: commenters should be respectful, you should use appropriate language, you cannot level personal insults at the author or at other commenters. Yet based on everything that we know about descriptive norms, it seems much more likely that the best way to encourage good behavior in a comments section is instead to model what good behavior looks like. This can involve stringently moderating comments sections so that disrespectful ones do not even appear, or including a note in the comment policy about how the author appreciates that "most people" contribute thoughtful, respectful comments.

It is also important to consider that the people with whom you are arguing in your comments are not the only ones affected by behavior in your comments section. As many others have noted, when you argue with a "troll" in your comments section, you are not just trying to persuade that person—you are also working to correct misinformation and persuade all of the silent readers observing the interaction.[2] This is no minor task, and your power here—and the potential negative power of your trolls, if left unaddressed—should not be understated.

Research has shown that hearing the same opinion stated over and over can lead people to overestimate the prevalence, popularity, and even accuracy of that statement, even if the repetitions are all actually coming from the same source.[3] Not only that, messages received from unreliable, untrustworthy sources can become more persuasive over time, as people forget where they heard the information (thereby forgetting to "discount" it as untrustworthy) and simply remember the content of the comments themselves.[4] It is very important, then, to model good behavior, correct misinformation, and address nasty commenters.

Nip Trolls in the Bud: The Science of Persuasion

One of the most fundamental models of persuasion research is Richard Petty's and John Cacioppo's Elaboration Likelihood Model, which states that people with varying levels of motivation or ability to attend to a message will take different routes to persuasion.[5]

Someone with a high level of motivation and/or ability—who finds the topic personally relevant, is knowledgeable in the domain being discussed, or is learning about the topic because she or he is responsible for some important relevant outcome—is likely to take what's known as the "central" route to persuasion. People taking this central route are more likely to pay attention to the arguments made within the message itself, and so will be persuaded by strong, high-quality arguments and dissuaded by weak arguments. Writers tend to focus on these readers, and on strong, higher-quality arguments, when trying to convince readers of their point of view.

By contrast, someone with a low level of motivation or ability—someone who is tired, distracted, or clicking between several Internet tabs—would be more likely to take the "peripheral" route. This involves persuasion through "heuristics," or automatic mental shortcuts that we often use to make decisions when we don't want to think too hard. Some classic examples include appealing to consensus ("four out of five doctors recommend this product"), using at-

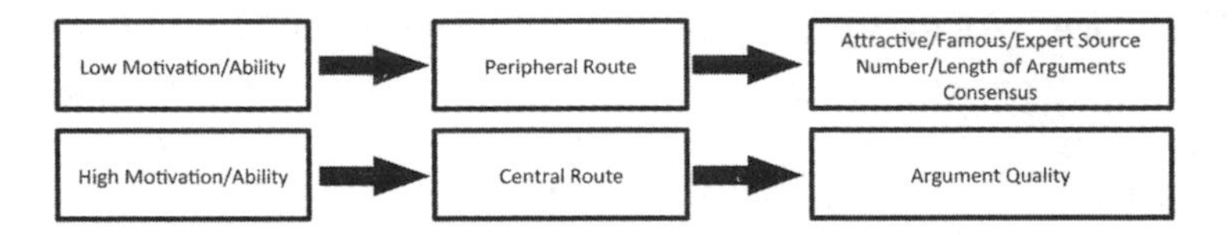

A basic outline of the Elaboration Likelihood Model, adapted from Richard E. Petty and John T. Cacioppo, "The Elaboration Likelihood Model of Persuasion," in *Advances in Experimental Social Psychology*, vol. 19 (Waltham, Mass.: Academic Press, 1986), 123–205.

tractive or famous message sources, or simply appearing to have a large number of long arguments.[6]

Broadly, what this means is that the same persuasive strategies will not work with every audience. We must acknowledge that many readers—especially on the Internet—are not particularly motivated to pay close attention to what we are writing, and they are often tired, distracted, multitasking, or otherwise splitting their attentional resources. This makes them more likely to be persuaded via the "peripheral" route. That is, if they see a message coming from a well-known celebrity, if it looks like there are a lot of arguments (but they don't necessarily read those arguments to see if they're actually good), or if they think that others tend to agree with the point of view being presented, they are more likely to accept the message itself, even if the actual arguments being presented are not particularly convincing.

You can use this model to your advantage in one of two ways. First, if you have high-quality arguments and you really do want your audience to pay attention to them, you must try and do everything you can to increase your audience's motivation and ability to pay attention to your message. And second, understand that many of your readers will likely be unmotivated and/or largely inattentive, and will opt for "peripheral" rather than "central" route strategies. As counterintuitive as it may seem for an inattentive audience, writing a long piece (that places your main argument in the headline or the lede) may actually work.

"Foot in the Door" and Cognitive Dissonance

Any good salesperson will tell you: if you're trying to get someone to agree to something big, start small and gradually raise the stakes. Once you get someone to agree to a small, easy request, he or she will be more likely to continue agreeing to successively larger requests. If you know that you're trying to convey a message that will be difficult to swallow, start small—don't plop the big message right there in the lede.

Why does this strategy work? Primarily because once people have started a pattern of "agreeing," it's hard to shift gears and start disagreeing—it's much easier to simply continue down the already established path of agreement. According to psychological research on cognitive dissonance, people hate seeming like "flip-floppers," whether that means acting differently from how they have acted in the past or behaving in ways that conflict with existing attitudes.

Is it ridiculous that changing your mind in the face of evidence could be interpreted as a weakness? Yup. But unfortunately, it's a hallmark of how our stubborn brains tend to process information. It's just easier for people to adopt a new opinion without actually "changing" their beliefs or sense of right and wrong.

Knowing this as a blogger means realizing that a better strategy than trying to change readers' minds or beliefs outright may be to find ways to make your point of view congruent with their existing convictions. Let's say that you are against the death penalty, and you are trying to convince someone else to adopt your opinion. If your opponent is arguing that the death penalty is right because it's "morally wrong" to make the taxpayer suffer the burden of paying for a criminal's life in prison, pointing out that the death penalty costs more is actually congruent with that conviction. (Calling attention to the Eighth Amendment and appealing to your opponent's sense of right and wrong by saying that the death penalty constitutes "torture," by contrast, would likely not be successful.) But if your opponent is inclined to care about the issue because he or she thinks that it is immoral to keep someone in prison for life because it's more humane to kill him quickly and swiftly—and that is why he or she supports the death penalty—then pointing out how executions themselves can be considered a form of torture (and your arguments for why this is the case) could very well be an effective strategy for persuading your opponent to shift his or her view. Any approach that makes it easier for someone to change his mind without actually appearing like he's really "changing his mind" will be more successful.

There is no one hard-and-fast way to guarantee that you will persuade every reader. And sadly, there will always be trolls. But by perpetuating descriptive (rather than prescriptive) norms about good commenter behavior, you can shift the tone of your comments section in a friendlier direction. In addition, by targeting the best types of arguments for your audience, easing any disagreeable readers into agreeing with you, and allowing readers to change their minds without really "changing" their minds at all, you can hopefully nip some of these bad behaviors in the bud.

Moreover, by invoking the aforementioned power of norm-setting, you can build a community that values active commenting, so you don't always have to be the one answering all of the comments on your own. And about communities—take a look at all of the contributors in this book. The science blogging community is an active, communal, and supportive one. It can sometimes be helpful psychologically to switch from an independent mindset to an interdependent one, especially if that will help you appreciate and learn how to rely on one of the best resources that you've got—your fellow bloggers. So stop. Breathe. Step away from the Internet. Hopefully with some of these tricks in mind, engaging with your cantankerous commenters does not have to be so awful—it might even become a new, fun kind of challenge. Just don't let the trolls get you down.

MELANIE TANNENBAUM is an award-winning social psychology instructor who taught at the University of Nevada, Reno. She blogs at *PsySociety*, which is part of the *Scientific American Blog Network*. Her writing has also been featured in *Scientific American*, *In-Mind Magazine*, and *The Open Lab Anthology: The Best Science Writing Online, 2012*.

Melanie is based in the San Francisco Bay area, California. Find her at her website, http://www.melanietannenbaum.com, or follow her on Twitter, @melanietbaum.

Notes

1. Robert B. Cialdini et al., "Managing Social Norms for Persuasive Impact," *Social Influence* 1, no. 1 (2006): 3–15.

2. David Shiffman, "Dawn Take You All: Bilbo Baggins' Approach Is Better than 'Don't Feed the Trolls,'" *Southern Fried Science*, May 21, 2014, http://www.southernfriedscience.com/?p=16389.

3. Kimberlee Weaver et al., "Inferring the Popularity of an Opinion from Its Familiarity: A Repetitive Voice Can Sound Like a Chorus," *Journal of Personality and Social Psychology* 92, no. 5 (2007): 821–833.

4. G. Tarcan Kumkale, "The Sleeper Effect in Persuasion: A Meta-Analytic Review," *Psychological Bulletin* 130, no. 1 (2004): 143–172.

5. Richard E. Petty and John T. Cacioppo, "The Elaboration Likelihood Model of Persuasion," *Advances in Experimental Social Psychology* 19 (1986): 123–205.

6. Understandably, the terms "central" and "peripheral" can be confusing, es-

pecially since in the diagram the routes themselves are both linear, not "centrally" or "peripherally" located in some way. These terms refer to the parts of the message that the recipient is paying attention to, not the position of the route itself. Participants taking the "central route" are paying attention to "central," fundamental aspects of the message; participants taking the "peripheral route" are paying attention to more tangential aspects.

25

Who's Paying?

Science Blogging and Money

BETHANY BROOKSHIRE

Online science communication might be fun, and it might teach the communicator a lot about the craft. But in the end, we all have to eat. Bethany Brookshire, a scientist who made the transition from part-time freelance blogger to full-time staff blogger at Science News, *will tell you how to make the blog pay off financially.*

When science writers and scientists first started writing science blogs, they were often framed as casual affairs. Some people still refer to their science blogs, video series, podcasts, and the like as "laboratories"—places where they can play with new ideas and new forms of presentation.

When I started blogging in 2008, I felt that my science blog was a volunteer effort. I was using it as a classroom, a way to improve my own writing and reach the masses at the same time. Many academics feel uncomfortable even asking for money; they assume that blogging falls under "outreach activities" and is therefore something they have a responsibility to do for free. It was considered enough to blog for "exposure," for a bigger audience, for more outreach.

Much has changed since 2008. Blog posts are increasingly formal, edited pieces, many of which appear on well-known sites. Bloggers do extensive reporting and fact-checking. In the absence of editors or other writing support, many bloggers end up doing it all, single-handedly producing polished pieces that are among some of the best science writing currently available. These days, the distinction between blogging and writing can be a matter of reporting, opinion, or editing effort—or simply be entirely arbitrary.

As my own blog grew in popularity, as I put in more time and effort, and as I began to see that this would be my career, I started to ask myself: why am I doing this for *free*? Now when people ask me if I can write a post for them, and say that I will benefit from "exposure," I have a standard reply: "People die from exposure."

In this chapter, we'll explore the different ways to use your blogging career to add to your pocketbook. Why? Because you are worth it.

Know Your Worth

When I was an academic blogging in my free time, it took me a long time to realize that my words were worth money. I bought into the idea that my efforts constituted outreach and that it was my responsibility. Perhaps it is. But that doesn't mean that it should be an uncompensated responsibility. For while blogging for free might be high-minded, it negates the efforts of the many professional science communicators out there. We work very hard, and deserve to be paid for what we do. Every time someone is willing to come in and do the same work for free, or for less, that's one less chance for professionals to make a living. This race to the bottom, with people always willing to write for a lower price, lowers the market value of our work. The less editors feel they have to pay, the less they will offer.

Your high-quality work, your time, is worth more than that.

Starting Out

As you start out, of course, no one will know who you are. You're not necessarily going to be able to "sell" your blog right off the bat. Building up a body of high-quality posts is the foundation for your money-making endeavors. Produce good content, and do it regularly. For some this may mean five days a week. For others, this may mean more extensive pieces once a week, or even once a month.

If you are already in science writing, blogging may be extra work in addition to a full-time writing job, or a supplement to a paying internship or freelancing for other outlets. It may also be extra work in addition to a full-time career as a scientist. But a professional body of work is essential to showing what you can do. Your blog or site can serve as a place not only to experiment with writing and to build up good posts, but also to collect your work—your clips, appearances, and so on—and show it off. It's a place for potential editors and employers to find more of your work.

Once you have content, you need visibility. The science communication community on the Internet is as much about networking as it is about communicating scientific ideas. Look for someone with the audience you want—maybe a lot of non-scientist Twitter followers. Connect with them and tell them about your work. Find other science communicators and ask for their feedback. Ask them for advice and get prominent people to promote you. Make and maintain a social media presence where you share not only your work, but also the work of other writers whom you admire, so your Twitter feed, Facebook, or other site will be a place where people go to find good content. Building contacts with other science communicators makes it more likely that editors will see your work. And people who know you and respect your work will make it their priority to hire you and pay you.

When I first started getting paid for my blogging, I was afraid to ask for too much. I was worried about meeting editors and was sure they would not be impressed if they met me. Finally I met someone with influence at a conference. As I blogged for the conference, I apparently impressed her. I made an effort to stay in touch with her, because I knew she hired writers. In the end, she encouraged me to apply for a large and very advantageous position that she had open. I had to produce a proposed budget and pay schedule along with my outline for the project. When I sent it in, she sent it back immediately, telling me to "look at the budget again." What she meant was "ask for more money." Never be afraid to ask.

Once you are writing with regularity, and have built a network of professional contacts, there are several roads to a paycheck in the world of blogging. Some may work for your personal site. Others may benefit both you and your career in science communications.

Ads

Google Ads and other ad software are some of the easiest things to add to your site. These simple widgets can be added to WordPress or blogger templates, or to your own custom-made template.

Ad software has immediate rewards. It doesn't require editors or pitching. All it requires is an audience. Keep in mind, however, that ads seldom bring in much money. Most ad software requires views in the millions to be profitable, and most science communicators who go this route are looking merely for enough cash to cover the costs of site hosting, without any extra for time and effort. Jason Thibeault, a blogger at *Freethought Blogs,* says of their Google Ads, "We make enough to keep the lights on and get a few bells and whistles now and again." But he notes that most of the bloggers at the site rarely get more than fifty dollars per month.

It is also important to keep in mind that with many types of ad software, you can't control what a viewer sees. It can be awkward to write a post condemning those who oppose genetically modified foods . . . and have an ad touting non-GMO products in the sidebar. Thibeault reports that at *Freethought Blogs* "we will often have to play whack-a-mole with objectionable ads, where we will ask our ad managers to remove certain ads, and they'll just reappear later when the ad campaign picks up a new buy." So while ads may be worthwhile to keep the site running, they do have their downsides.

Other Sponsorships

For those who do end up with a wildly popular blog, Facebook page, or YouTube channel, full-time jobs may be available. Your YouTube channel might be picked up by PBS or another large group, or companies may request to advertise on your page.

These jobs, sponsorships, and ads can be useful and even lucrative, allowing you to devote yourself to those projects full-time. But they should also be considered carefully. Will having this particular sponsorship add or detract from your own reliability as a communicator? A video series supported by PBS is a good thing. An energy blog supported by a multinational oil corporation might be more suspect. If you are a freelancer, it's also important to read contracts. What would the sponsorship mean for your content? Does your sponsor want editorial control? Who has the rights to your work?

Grants

There are grants out there to help science communicators achieve their goals and get their work out to the world. Many of these are intended for dedicated sites on a single topic, or for book projects. The National Association of Science Writers, for example, funds grants that focus on science communication, including books like this one. The Alfred P. Sloan Foundation also offers grants for science communication work. Many universities offer science communication scholarships, and some other organizations offer short-term fellowships. There are many other outlets that might help the budding science communicator get paid for his or her content (some of which are listed on the companion website for this book).

Grants, however, are never a sure thing. Many have a relatively narrow focus or are extremely competitive. Day to day, it might be best to rely instead on one of the following strategies.

Try a Network

There are many reasons, personal and professional, that joining a blogging network may or may not work for you. But most of the larger networks, such as *Scientific American Blog Network, National Geographic Phenomena, Discover Magazine Blogs,* and *Wired* do pay their bloggers. Dollar amounts vary, and are almost never enough by themselves to make a living, but they do provide some compensation. And they connect you with a well-respected, established brand.

The byline connected with your writing can, in turn, bring in more paying work. An invitation to join the *Scientific American Blog Network* in 2010 allowed me to use the "SciAm" name at conferences. This led to better networking, meetings with editors, and freelancing opportunities at the *Guardian* and *Slate*. Eventually, having been on a high-profile network like *Scientific American*'s led another magazine to offer me a position as a full-time staff writer. The paycheck from my blogging may have been small, but the professional dividends were considerably larger. In the end, they led to a steady salary.

If you are going to try a network, it's worth reaching out to editors and writers who are already part of that community. Ask what the requirements are (a certain number of page views or a certain number of posts per month, for example), what the experience is like, and how much editorial control they have. Ask writers within those networks if they are willing to say how much they get paid and to describe the basic terms of their contracts.

If you reach out to editors, ask what they are looking for in a blogger. Is there a niche they are looking to fill? A new point of view? A different format like video or podcasting? Read the collection of blogs already present: is there a space for yours? Could you offer a new perspective or attract a new audience? If you find a space where your blog can fit, you might have an easier time selling your blog to a network. Ideally the network will benefit from your presence while you benefit from its name and support.

Freelance Blogging

If you are a science writer setting out to make a name for yourself, freelance blogging can help to fill the gaps between larger pieces. Not only does freelance blogging get your name out to a wider audience; it helps you to meet editors, build your clips, and even make a little bit of cash.

Many outlets now have paid freelance bloggers, hired on a post-by-post basis. *Slate, Scientific American, Discover, io9,* the *Guardian,* the *BBC,* and *DoubleXScience* are just a few examples of sites that hire freelance science bloggers. While such outlets tend to pay much less for blog posts than for full pieces, blog posts also tend to be assigned faster, come out more quickly, and go through a less extensive editing process.

Search for science sites that publish pieces in your area of expertise: do they hire freelancers? Are any of those freelancers also bloggers? What kind of bloggers do they hire and how might you fit that profile? How do the blogs on a given site differ from the other content?

Reach out to those who have previously blogged for your site of interest. Ask them about their experiences, what the process was like, and how they pitched their ideas. Ask them how much they expected to get paid. Not everyone will tell you, but it doesn't hurt to ask.

As you make your way around the world of science communication, both at conferences and on social media, keep an eye out for editors. It helps to introduce yourself to editors and make them aware of your work. Ask them what they are looking for. When the time comes, send them a pitch.

When I was beginning to write professionally, I was lucky enough to meet Laura Helmuth, the science editor at *Slate.* As it happened, she was looking for writers, and was especially interested in promoting female science writers on the site. When an editor requests pitches, don't lower your eyes modestly and say no. I pitched. I got a post. And I got paid!

Writing a pitch for a blog post is very much like writing any other pitch for any other type of piece. Consider what makes your proposal novel. Think about how it fits into the venue. But also consider how much time you are willing to put in to write it. Many editors use blog posts as a mechanism for getting more content for less money. You will probably get paid less—perhaps as much as 90 percent less—than for an "article" of the same length.

This may make you consider your options. Can you offer the post to a network that pays better? In addition, consider how much time you may want to offer. No one would want to turn in bad work, but at the same time, a sixty-dollar blog post may result in less perfect prose than your first five-dollar-per-word *National Geographic* feature. You get what you pay for, after all.

Odds and Making Ends Meet

Blogging on its own may not routinely bring home the bacon. But the experience and networking that you gain from a social media presence has other benefits. Many bloggers, myself included, have gone on to paid speaking gigs, freelance writing and editing assignments, and paid podcasting and video projects. Many bloggers now make their way as freelance writers. My blogging finally paid off in a full-time staff position.

What can you conclude from all this? There is no one way for a science communicator to make money. There are networks, Google Ads, sponsorships, blog posts, videos, and audio for various outlets. Some outlets pay well, some pay less, and some will ask you to blog simply for exposure. As you progress in your career, weigh your options carefully and keep track of what you hear as you build your professional network of other communicators. Know which sites will pay, how well they pay, and what kind of work they are most likely to publish. In the end, freelance blogging isn't very different from freelance writing more generally, as blog streams and freelance writing rivers meld into one digital writing ocean. This means that just like with freelance writing, you should never be afraid to ask for fair compensation. You are worth it.

BETHANY BROOKSHIRE is an award-winning science writer at *Science News* and *Society for Science & the Public.* She runs the *Scicurious* blog and manages social media for *Science News for Students.* She is also the guest editor of *The Open Lab Anthology: The Best Science Writing Online, 2009.*

Bethany is based in Washington, D.C. Find her at her blog, http://www.scicurious.org, or follow her on Twitter, @scicurious.

26

From Science Blog to Book

BRIAN SWITEK

Taking the first step from short form to long form can be daunting, but no one is more familiar with the transition from science blogger to book author than Brian Switek, author of My Beloved Brontosaurus *and* Written in Stone*. Here he shares his experience on how he transitioned between the two media, and how you can use a blog to pitch, write, and promote your dead-tree publication.*

There are no rules for what a science blog should be. A new blog post is functionally no different from a blank piece of paper. You're free to create within the bounds of what the form can hold. And because blogs are media platforms, they do not require any specific style or type of content. These attributes give blogs a unique versatility, and make them useful testing grounds for other types of media.

For science writers who want to compose a book, whether they're veterans with a new project or prospective authors trying to sell their first title, blogs can be powerful tools throughout the publication process.

Before Signing

Science books are sold on the strength of proposals. This is the most difficult and painstaking part of the entire process, especially for first-time authors. A prospective author has to identify an idea worth a book-length treatment, articulate that kernel of an idea with a detailed overview of what the book is going to be about, use that proposal to find an agent, and then refine it to catch the attention of a publisher. In essence, the proposal is an overview of your book before you've even fully figured it out. This means that the book you propose and the one you eventually publish will not be exactly the same. But that's expected. The point of this process is that a proposal will lay out the narrative arc of your book and help you figure out which parts of the story go where. The standard proposal involves an overview summary of two to five pages, a section on competing titles and why your book is different, and a chapter-by-chapter breakdown that briefly describes what each chapter is going to contain. Writer's Relief breaks down the section-by-section elements of what a compelling proposal must include.[1]

A tight and polished proposal will help lure an agent, and an agent will help you further improve the proposal before taking it to publishers. Agents are an essential part of today's publishing ecosystem—most popular science publishers will respond only to proposals that come through an agent. And if you do get an offer, a good agent will be able to haggle to provide you with a better deal and cut through the legalese of a book contract.

While most of the proposal-writing process occurs behind the scenes, science blogs can provide an open space for experimentation during this early stage of the book process. As science writer Jennifer Ouellette once said, blogs are writing laboratories. They are places to try out different styles, structures, and ideas with the added benefit of feedback from readers in the form of comments and traffic. Finding a concept that can support a book and not a magazine story or long-form feature requires writing down certain storylines to see if they open up or crumble. Blogs are a handy proving ground for these early stage book concepts. And since blogs are easy to archive and search, authors can readily go back to review ideas or stories that take on new relevance as projects proceed.

Blogs also offer writers the ability to refine their tone, be it no-nonsense (http://phenomena.nationalgeographic.com/blog/the-loom) or playful (http://blogs.scientificamerican.com/running-ponies). After all, books aren't just about content; prospective authors need to consider issues of style and perspective as well. Some books might benefit from—or even require—the author's involvement in the story, while others are best told from the third person. Blogs allow authors to find the voice they want to present in book form.

Reader reactions can help guide readers through these choices. If a post gets an unexpected response from readers or generates a surprising amount of interest overall, aspiring authors can look into that topic to see if it might fit into a larger pattern worthy of a book. Gauging public interest from an early stage can help refine a book's central question or thesis into something that the average reader will be interested in rather than a topic of interest only to specialists. The trick is to make sure that your blog plugs into the same audience as you hope to reach with your book. If you write for physicists, and are principally getting feedback from physicists, then your idea probably isn't going to translate well for those who aren't as invested in Newton and Einstein. Then again, if your aim is to reach such a niche audience, building that sort of specialized readership can provide a useful sounding board during your book's development. You will have to be very intentional about your audience throughout the project.

But writing a science book requires more than an original idea. Publishers look for authors who have a broad social media reach and are recognized as prominent voices on science. While not a pleasant truth, marketing concerns can sometimes make or break a deal. Blogs can help overcome this obstacle for writers, especially early career writers who do not yet have a long list of varied clips at hand. Blogs allow writers to show off their expertise in their chosen area of science, connect with experts who could be helpful later in the writing process, and draw the attention of editors who may be able to offer freelance work to build a list of clips.

Ultimately, your blog traffic, the venue you blog at (be it independent or a network), and the way you've developed a personal brand will all play into how you are able to catch the interest of an agent and an editor. If you're serious about writing a science book, the earlier you can start blogging and distinguishing yourself as a unique voice, the better.

While Writing Your Book

So you've successfully navigated the hazardous terrain on the way to signing a book deal. Don't stop blogging. For one thing, readers require regular care and feeding. If you hope to carry your audience along with you, you'll have to maintain a regular online presence.

If used strategically, science blogs can also allow authors to research their book while building their public exposure. Just as science blogs can be helpful in refining big-picture book ideas, they can also be used effectively to gather anecdotes and other material for the larger work. Science does not wait for writers. Each week there is a new spate of fresh research in a variety of journals, some of which may be relevant to the book at hand. A science blog, which requires regular posts anyway, provides authors with a platform to discuss new finds related to their topic and explore whether those should go into their book or be left out. By keeping up with and writing about studies related to their project, authors will both stay current and bolster their reputation as a good source of information and analysis on their chosen subject.

The trick is to avoid repeating yourself. While it is perfectly fine to write in both your blog and your book about research in your field, you won't want your editor—much less future readers—to look at your book and discover a series of recycled blog posts. So unless an editor has specifically commissioned a book of blog posts, simply copying and pasting blog posts—or even parts of blog posts—into a book manuscript is a terrible idea. Recycling old text not only may run afoul of contractual obligations to write new material; it will also reduce the incentive for readers to pick up your book. Instead, when covering topics in your book that you've previously blogged about, start fresh. Doing so will lessen the chances of you repeating yourself and will ensure that your points flow organically with the rest of the new text.

While blogging about stories related to your book can be tricky, doing so will become more worthwhile as your book starts to coalesce. The ideas and studies that are mentioned in the book will change throughout the writing process, all the way to the end. Especially as the final manuscript starts to form, blogs can act as useful databases of what you've previously written about. Let's say that you're down to the last edit, and you remember a study relevant to your final chapter but you can't remember the citation. You know you blogged about it, though, and with a quick search you can pull up the post and check to see if a discussion of the study might fit into the book.

Using a science blog to effectively make public notes on ideas relevant to your book has another advantage. If from the start of your book manuscript's development you have monitored how readers have responded to subjects that you wanted to include, you'll be tuned in to what strikes your readership as especially interesting. This means that when your book is completed, you'll have a sense of which topics in it will likely grab the attention of readers and which will not—and you can use this information to refine your marketing message. After all, you want readers to look forward to your book! If done strategically, the effort of maintaining a science blog will be its own form of self-promotion that starts generating buzz for you as the book nears publication.

After Publication

There's no reason to stop blogging when your book comes out. If anything, authors should plan blog posts that tie into the book around the time of publication. The easiest way to do this is to set up a series of posts on topics that either didn't make it into the book or that had to be cut despite being interesting in their own right. Science blogs can be homes for book B-sides. Tying those posts into your book is simple enough to do, as easy as adding a line such as "If you'd like to know more, read . . ." with a link to your book.

Just as you shouldn't straight-up recycle blog material into your book, it is best to avoid publishing long excerpts of your book on the blog. With any luck, magazines and other short-form media will approach your publisher about running excerpts from your book; sharing those snippets on your blog first may scuttle opportunities to have them featured in other venues.

And there's no reason that all your promotional efforts have to appear on your own blog. Some blog networks and other science-related sites take guest posts. Some of these pay, and some do not, but authors should identify other blog venues that might consider a submission. Even a relatively short post sharing a small story or idea from the book can gain the attention of an audience that you may not have been able to reach otherwise. Similarly, don't be afraid to ask other bloggers—preferably those with a larger following than you—to mention your book's publication or to run an interview with you. Use the blog and social media connections you've made since the very start of the project to help get the word out through reviews and tweets.

Every book project will be as different as the authors who write them. Veteran writers with a few books behind them will obviously have different needs than a scientist who wants to write her first book or an early career journalist who wants to pursue a question too big for a magazine. But no matter your background or career stage, blogs are excellent tools for developing books and will help you to have a prominent voice in the science communication realm. Don't consider blogs to be separate from the other writing involved in a book project. Along with other forms of social media like Twitter and possible freelance pieces, blogs can reinforce and amplify an author's signal in a crowded media landscape. The path from blog to book can be frustrating, and even treacherous, but when employed carefully, social media can help you to navigate it successfully.

BRIAN SWITEK is the writer of two popular science books on paleontology, *Written in Stone* and *My Beloved Brontosaurus*. His blog is hosted by *National Geographic*. He has also written for *National Geographic*, *Slate*, and *Nature*, among other publications; is the science writer for the film *Jurassic World*; and is the host of dinologue .com's video series *Dinovision*.

Brian is based in Salt Lake City, Utah. Find him at his website, http://www.brianswitek.com, or follow him on Twitter, @Laelaps.

Note

1. Writer's Relief, "Writer Wednesday: How to Write a Non-Fiction Book Proposal," *Huffington Post,* July 10, 2013, http://www.huffingtonpost.com/2013/07/10/nonfiction-book-proposal_n_3569043.html.

Afterword

PAIGE JARREAU

In September 2014, I wrote a blog post listing over a hundred women scientists' Twitter handles in reaction to a contentious article in the journal *Science* titled "The Top 50 Science Stars of Twitter."[1] As soon as it was published, *Science*'s article, itself a critical response to Neil Hall's "Kardashian Index," had prompted a storm of disapproving tweets, many using the hashtag #WomenTweet ScienceToo, and blog posts.[2] The "top 50 science stars of Twitter," it turns out, were overwhelmingly Caucasian and male. But the hundred women scientists were not alone: a large number of scientists and science communicators, both men and women, found *Science*'s list to be egregiously biased. Many, including myself, took to blogging not only to say so, but also to provide alternative lists and alternative ways of determining who the real science stars of Twitter were.

Even five years ago, the blog-based conversations around *Science*'s original list would likely have remained within the domain in which they began: the blogosphere. But in 2014, the dynamics and influence of science blogging are quite different from what they were then. Just a few weeks after I published my post "In Response to the Top 50 Science List" on scilogs.com, *Science*'s deputy news editor John Travis linked to it in a sequel to the original news article.[3] In addition, the sequel's revised and expanded list of top "science stars of Twitter" featured twenty women scientists compared to the original list's four—all of whom, and many more, had been included in the list I had compiled in response.

This story is meaningful for two reasons. First, it was the first time that my relatively small science blog received the attention of, and importantly a link from, a prestigious scientific publication. The attention from *Science* was followed by a phone call from a reporter at the journal *Nature,* revealing that *Science* was not the only "big name" publication to follow the outrage in the blogosphere. Second, and more importantly, this story is meaningful because it is far from unique. Instead, it points to a broad increase in the influence of science blogs today. For science blogs are no longer relegated to critiquing science journalism from a niche corner of the Internet.[4] Rather they are increasingly becoming integral components of science journalism and the science journalist's work, even if they still serve their traditional roles of science media debunking, criticism, and community building among scientists. In short, science blogging has moved to the mainstream. And with this move has come science blogging's professionalization—as well as growing professional responsibilities for science bloggers, especially those wishing to break into the big leagues.

Individual science blogs may never reach the readership levels of traditional science news publications, but they may *surpass* them in their importance to the progress of science journalism and to the growth and diversity of the science media ecosystem.[5] Science bloggers are carefully choosing what they cover on their blogs according to shared goals of adding value, advancing the conversation, and writing what's missing from science news coverage. Individual bloggers' strategic choices to deep dive into the underreported science stories of the week are not only changing which stories are told, but also having far-reaching implications for science blogging itself.[6] The shared goals of science blogging have led to an explosion in alternative coverage of science that has spread far and wide beyond the limited scientific topics and information sources featured in traditional news outlets. Science bloggers have decided they can do much more than play second fiddle to mainstream science journalism, and their efforts are paying off. Nearly every science blogger I have talked to recently has seen one or more of their blog posts picked up by a mainstream news outlet. Science bloggers today are serving as expert sources to traditional news stories, if not breaking stories themselves.

My research on science-blogging practices and values, while yet in its early stages, has yielded many insights that are key to understanding the nature, impact, and promise of science blogging. If you are reading this book, you are likely interested in starting a science blog yourself or improving one you already maintain. At the very least, you are interested in the phenomenon that is science blogging. So one of the most important lessons I can leave you with is that currently, across the sciences, and in every realm of scientific engagement—from bench and field scientists to professors and teachers, from fresh-faced graduate students to funding agencies, and every role and job in between—those participating in science today take science blogs seriously.

As traditional journalists, scientists, and readers alike begin to take blogging more seriously, science blog authors are realizing that they need to be more professional in their blogging practices. Transparency, accountability, and fact-checking are all standards that today sharply delineate the blogs we pay attention to, read, and promote. Sloppy, ranting, or uncareful prose just isn't cutting it anymore.

Blogging networks have become communities of science writers who not only bounce ideas off one another, but also send each other drafts of blog posts for editing and fact-checking. We've moved away from the phenomenon of blog posts as back-and-forth online discussion forums to blog posts as stand-alone, rigorous, in-depth, contextualized, even long-form journalistic and editorial pieces. This doesn't mean that science blogs are any less interactive and discussion-oriented than they used to be, even if much of the "quick and dirty" discussions have moved to Twitter and other social networks. But it does mean that science blogs are being held to a higher standard. Clean interfaces and visual storytelling are also of rising importance to science blogging success, if we measure success by reader and media attention.

The professionalization of what we might call science "blogo-journalism"—an extension of rigorous forms of research blogging—is steadily raising the bar for what blogging about scientific research can look like.[7] Professional science bloggers such as Ed Yong at *National Geographic* and Bethany Brookshire at *ScienceNews* have become pioneers in creating rigorous, fascinating, and beautifully written blog posts on breaking research. And yet journalists, scientists, and users alike are also taking other approaches to science blogging more seriously than ever before. From science outreach, to education, to public engagement, to strategic communication, to advocacy, to open peer review, to editorial writing, to scientific networking and collaboration, it seems that our current media ecosystem often looks to science bloggers as the first line of voices on important issues in the public and scientific spheres. Science bloggers today are our public intellectuals, our media revolutionaries, our scientific critics, our watchdogs, our specialized journalists. Not surprisingly, then, the reputation a science writer develops through blogging is increasingly following that writer wherever he or she goes on the web. And we expect that writer to act ethically, to write responsibly, to tell us what we haven't considered, and to get it right.

So what does the changing nature of science blogging mean to you, as you continue developing your skills in the craft? It means that many of the promising aspects of science blogging you have read about here, from getting paid, to getting interactive, to measuring your impact, should be understood in terms of an increasingly professional and evidence-based approach to science blogging. Because today high-quality science blogging is much more than just blogging. It's journalism. It's watch-dogging. It's public engagement. It's open peer review. It's science in progress. And the beautiful part is that your personal experience and expert opinion are more important than ever.

PAIGE JARREAU is a science blogger and blog manager at SciLogs.com, where she authors the blog *From the Lab Bench.* She has a Ph.D. from the Manship School of Mass Communication, where she studied science-blogging practices for her dissertation. She has been named the 2015–16 Lamar Visiting Scholar at the Manship School of Louisiana State University, where she will begin her postdoctoral work in science communication.

Paige is based in Baton Rouge, Louisiana. Find her blog at http://www.scilogs.com/from_the_lab_bench, or follow her on Twitter, @FromTheLabBench.

Notes

1. Jia You, "The Top 50 Science Stars of Twitter," *Science,* September 17, 2014, http://news.sciencemag.org/scientific-community/2014/09/top-50-science-stars-twitter#full-list.

2. Paige Brown Jarreau, "In Response to the Top 50 Science List," *From the Lab*

Bench (blog), SciLogs, September 18, 2014, http://www.scilogs.com/from_the_lab_bench/in-response-to-the-top-50-science-list.

3. John Travis, "Twitter's Science Stars, The Sequel," *Science,* October 6, 2014, http://news.sciencemag.org/scientific-community/2014/10/twitters-science-stars-sequel.

4. Geoff Brumfiel, "Science Journalism: Supplanting the Old Media?," *Nature* 458 (March 2009): 274–277.

5. Paige Brown, "An Explosion of Alternatives," *EMBO Reports* 15, no. 8 (2014): 827–832; doi: 10.15252/embr.201439130.

6. Paige Brown Jarreau, "Do Science Bloggers Blog about What's Popular? Or Not?," *From the Lab Bench* (blog), SciLogs, June 5, 2014, http://www.scilogs.com/from_the_lab_bench/do-science-bloggers-blog-about-whats-popular-or-not.

7. Cornelius Puschmann and Merja Mahrt, "Scholarly Blogging: A New Form of Publishing or Science Journalism 2.0?," *Science and the Internet* (2012): 171–181.

Acknowledgments

No book gets written without the efforts and support of many, especially a book with as many contributors as this one. First we must thank all of our authors for their hard work, thoughtful contributions, and valuable insights. Special thanks as well to Paige Brown Jarreau for her concluding remarks. Jeanette Kazmierczak was of tremendous assistance as the final manuscript came together; her formatting help was invaluable. Thanks also to Joe Calamia of Yale University Press for his guidance and cheerleading throughout the long process in which a book like this goes from an idea to the final product, to Julie Carlson for making sense of twenty-eight contributors' worth of prose, and to Bora Zivkovic for helping to get this project off the ground. Finally, we are deeply grateful to Siri Carpenter and Jeanne Erdmann of The Open Notebook for working with us to put together this book's incredible accompanying website (http://www.theopennotebook.com/science-blogging-guide), where you will find interviews with chapter authors, live links to the references in this book, and many other nuggets of useful information.

Index

Page numbers in *italics* refer to illustrations.